FUNCTIONAL PACKAGING PROTOTYPES

Jinming chen

FUNCTIONAL PACKAGING PROTOTYPES

Published in the United States of America by Gingko Press
Gingko Press, Inc.
5768 Paradise Drive, Suite J
Corte Madera, CA 94925, USA

Phone (415) 924 9615
Fax (415) 924 9608

email: books@gingkopress.com
www.gingkopress.com

First published 2007 by Liaoning Science and Technology Press
Editor: Wang Lixia
Author: Chen Jinming

ISBN: 1-58423-268-4
ISBN-13: 978-1-58423-268-1

Printed in China

Contents

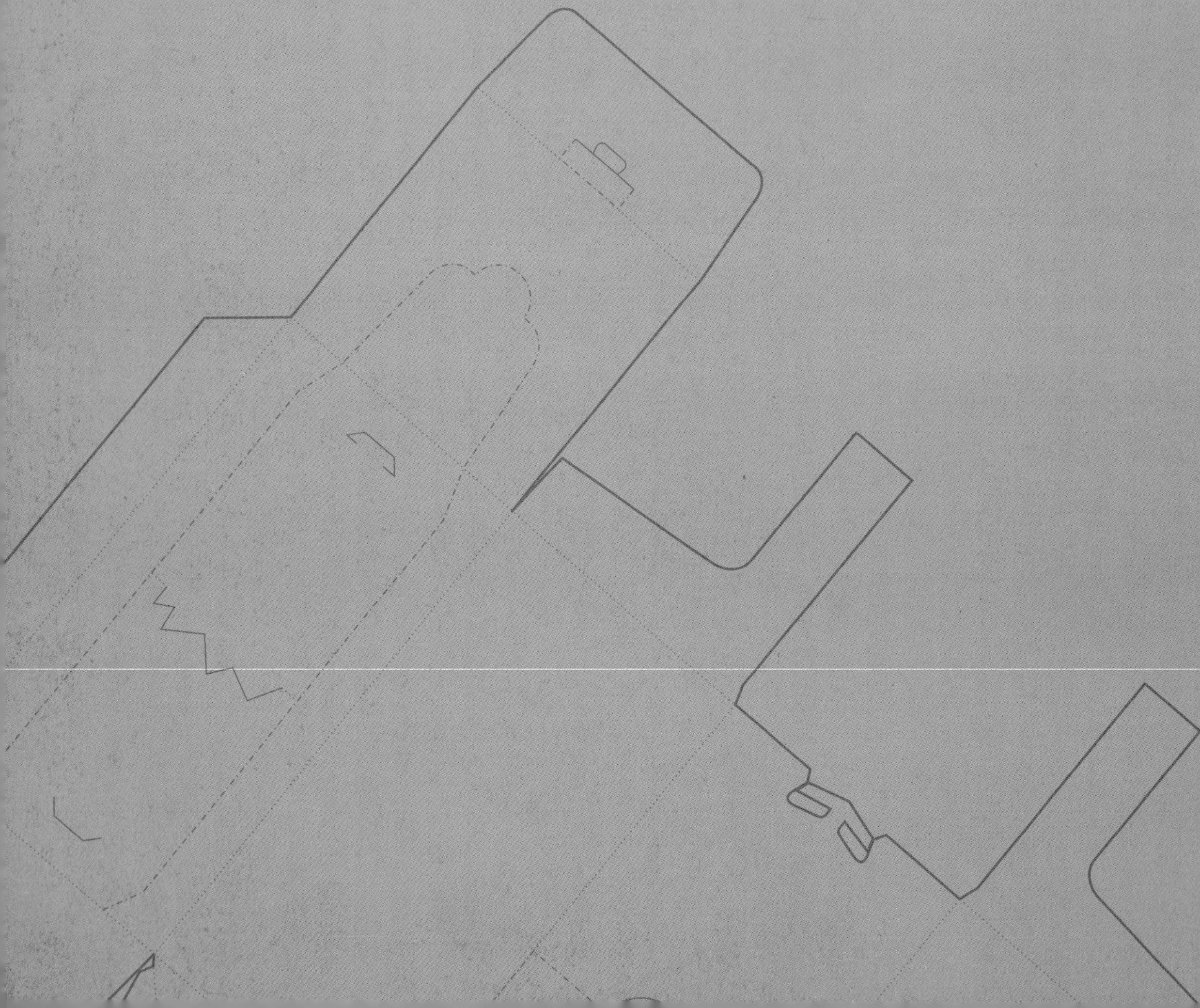

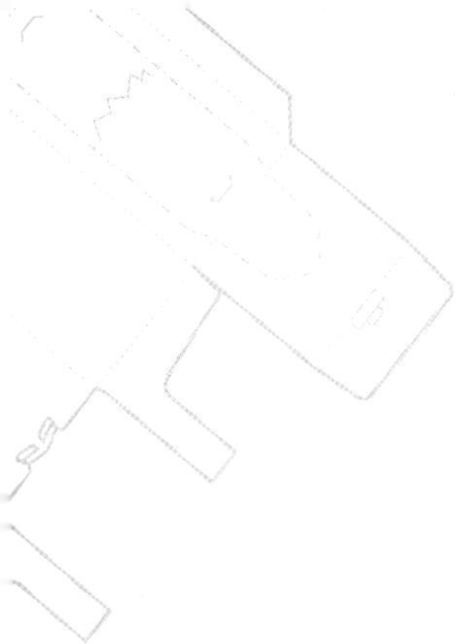

Foreword

This book collects 150 examples of the most creative packaging solutions selected according to their uniqueness, feasibility of manufacture and merchandising practicability. These innovative designs are illuminated using both photographs of the packages and structural templates, or blueprints. The templates are found on the enclosed CD-Rom and may be modified for the reader's own purposes.

诗蒂
Shinetien
诗蒂
Shinetien

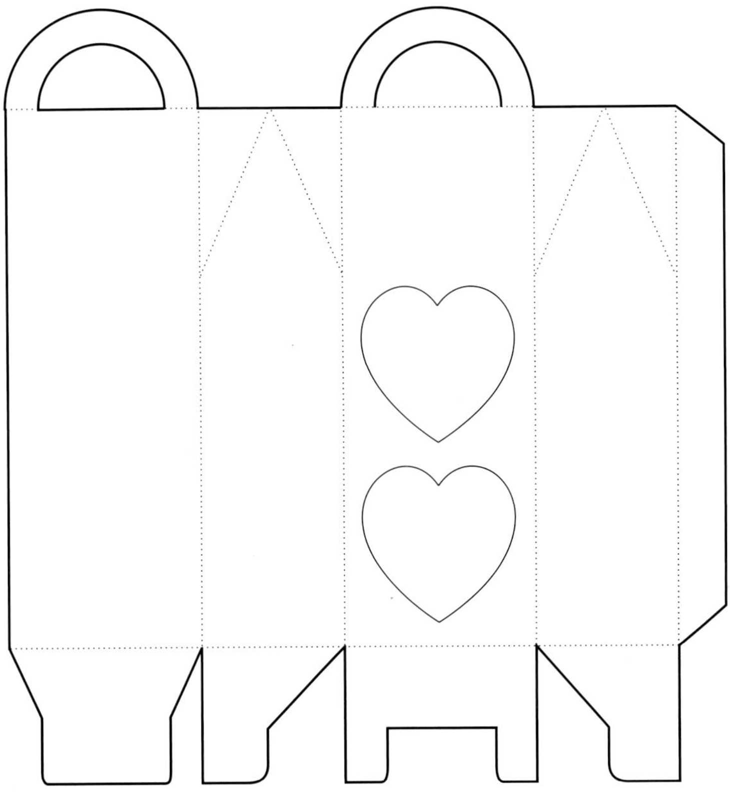

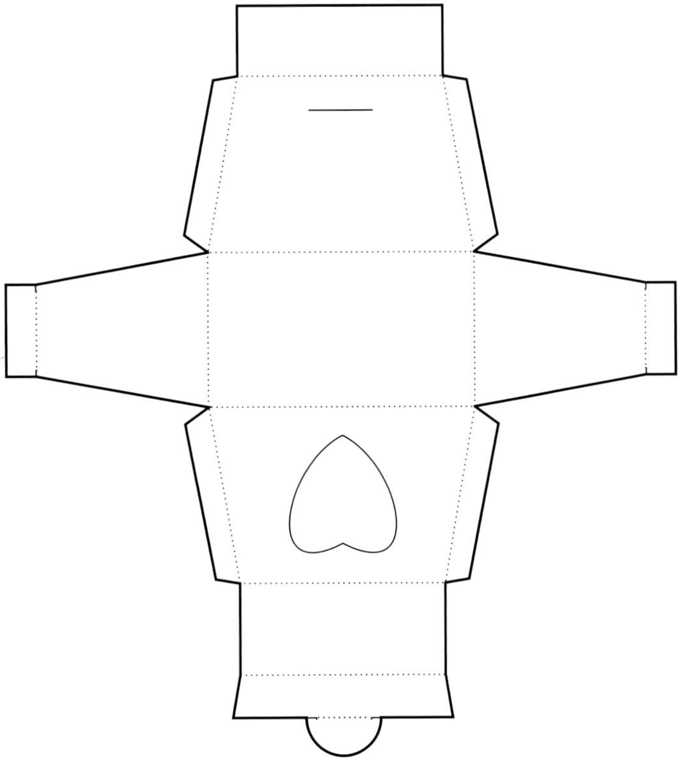

ShineTien
ShineTien Chocolate

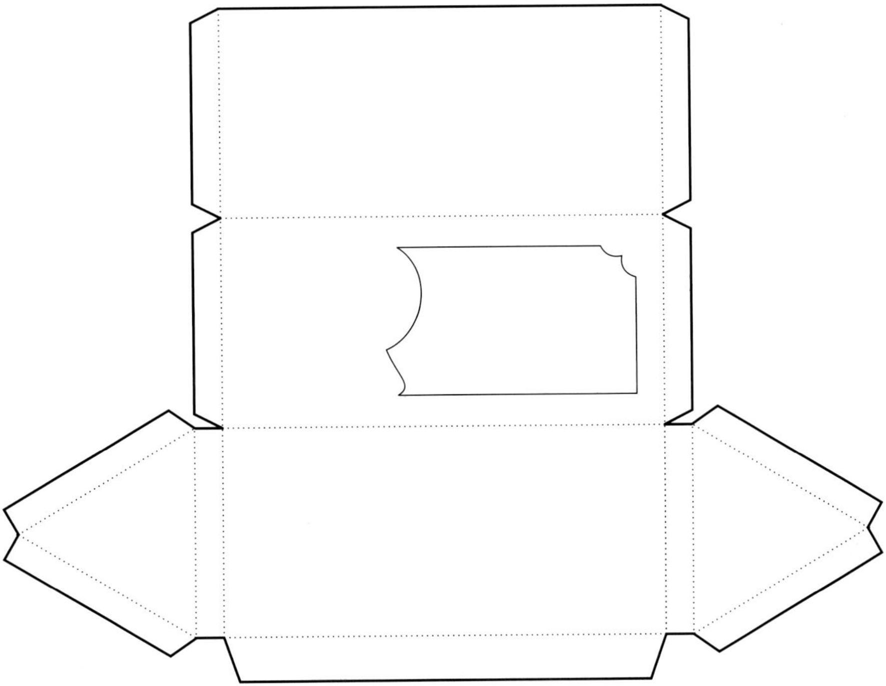

GuyLian
Belgian Chocolate
Twists
GuyLian
GuyLian
GuyLian
Mixed Selection
Filled Belgian Chocolates
NET WT/CONT NET 43g ℮ 1.51oz

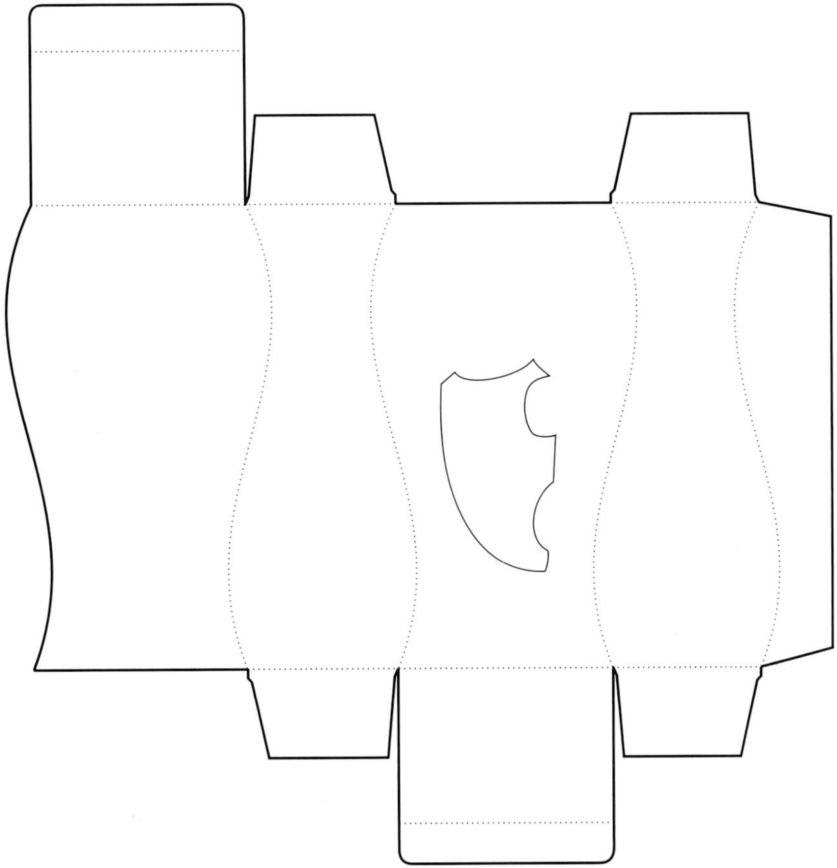

麗新語
Elleair Friend
高级化妆棉
Pure Cotton
素肌にやさしい
不織布と中棉なしつかりとシール
两面为化妆和卸妆而设计的专用棉
特别为化妆和卸妆而设计的专用棉
80枚入
Elleair Friend
合格C5110E023
生产日期20051108

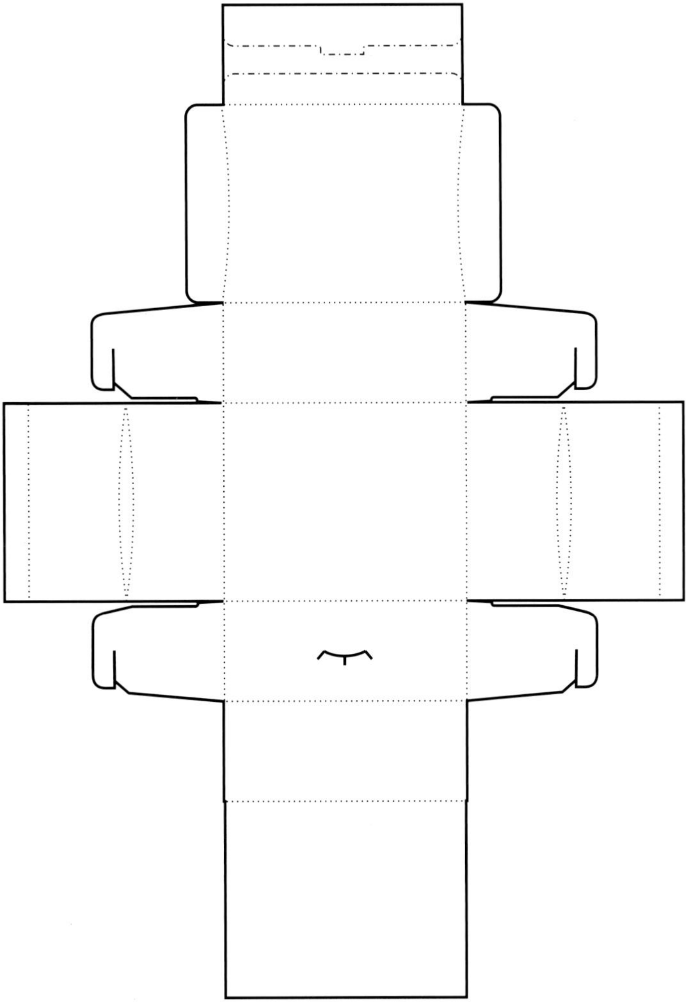

菠萝派
pineapple pie
CAUTION
HANDLE WITH CARE
I'M HOT
我就喜欢™
...i'm lovin' it®

LANCÔME
PARIS
BLANC
EXPERT
Mela-NO Cx
Advanced Whitening Light Blooming
Synergy Treatment
Traitement Duo Éclaircissant Sur-Activé
Révélateur de Lumière
4 masks x 23 ml / 0.7 FL.OZ ℮ 92 ml / 2.8 FL. OZ
4 bottles x 7 ml / 0.23 FL.OZ ℮ 28 ml / 0.9 FL. OZ

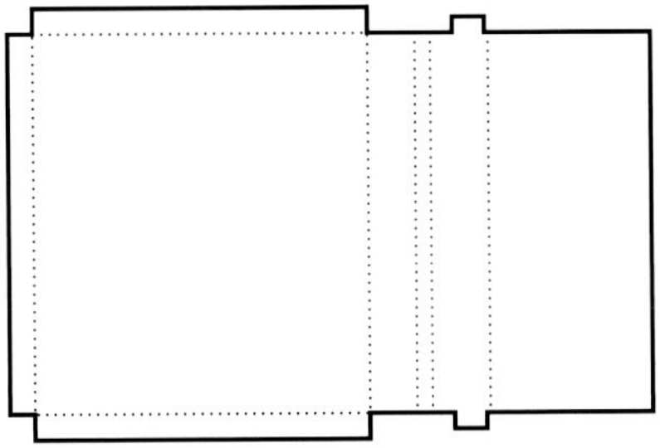

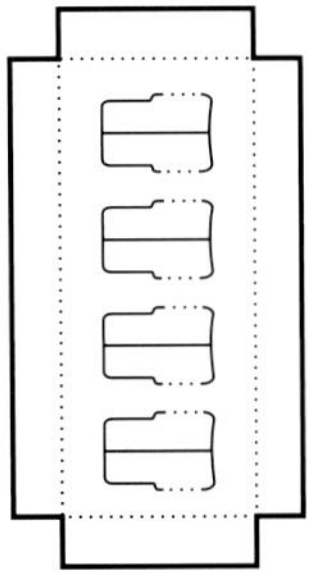

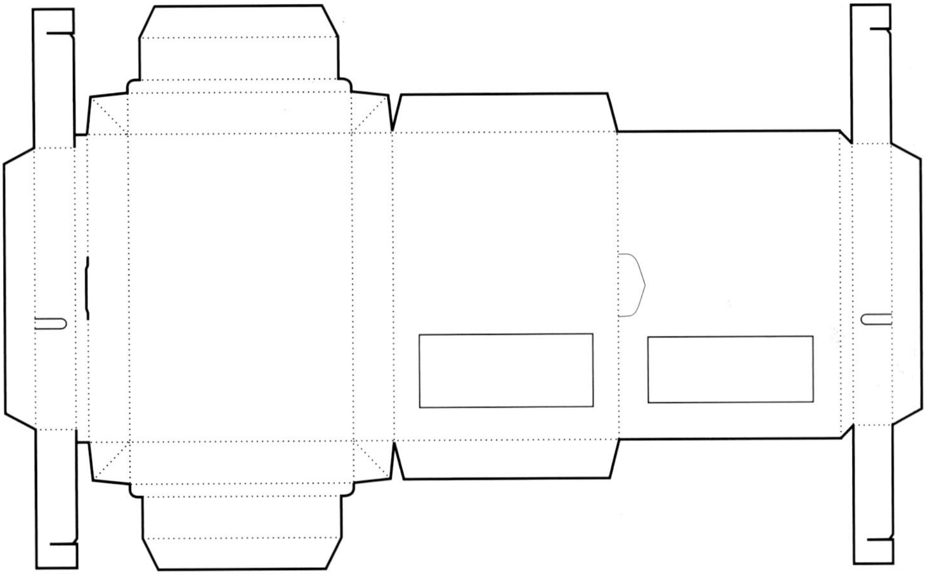

Nexcare™
耐适康
FIRST AID
您的随身小护士
隐形痘痘贴
3M
让痘痘没有明天!
综合装

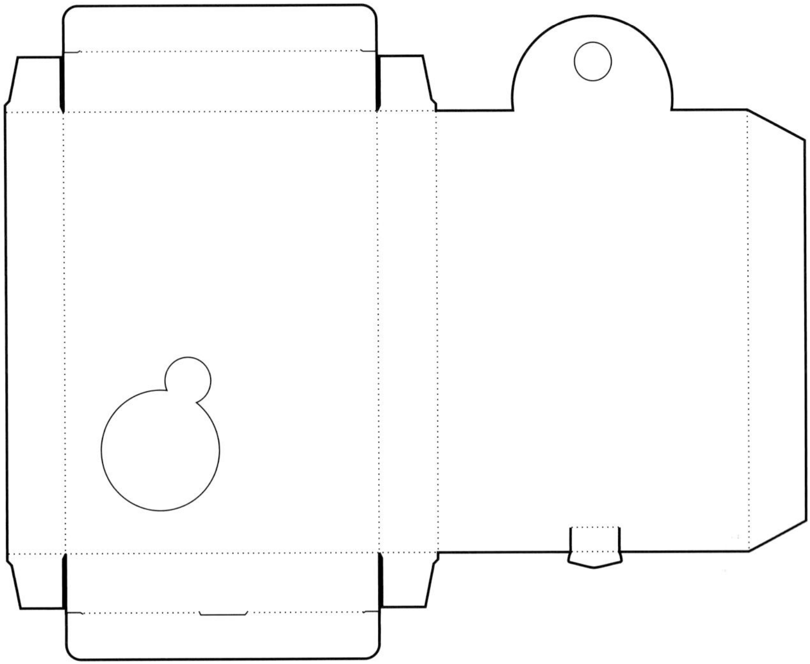

AVON

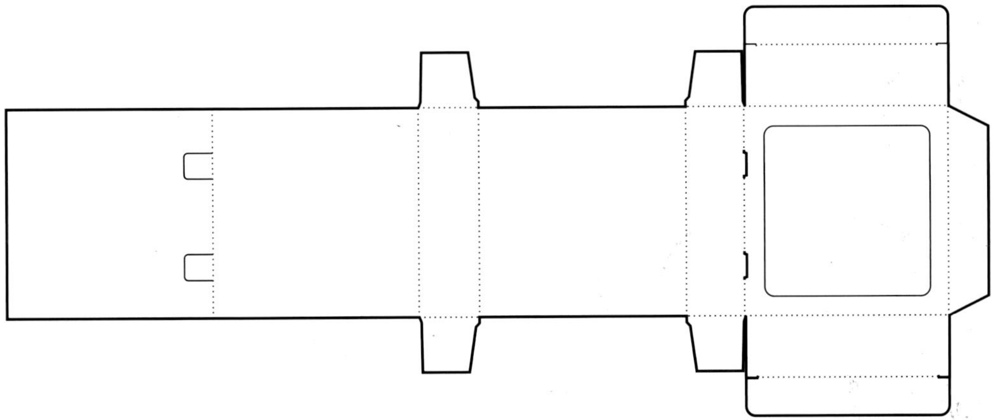

neo
Quality Bottle
Complete With Silicone Nipple & Leakproof Hood
优质塑料奶瓶
9oz
250ml
PIGEON

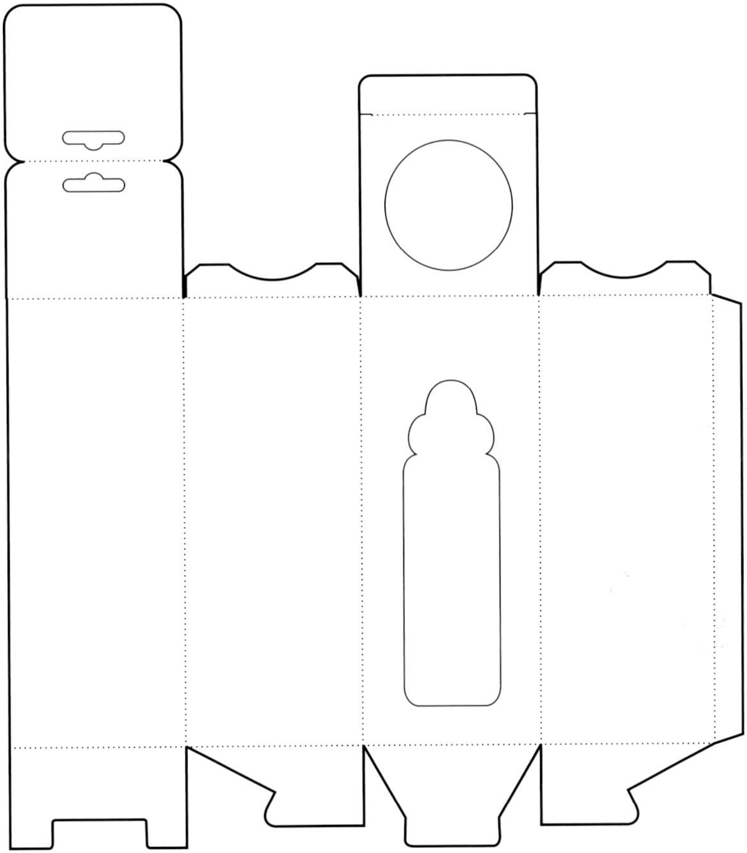

LUX
送 50ml
闪亮冰爽美肤沐浴乳

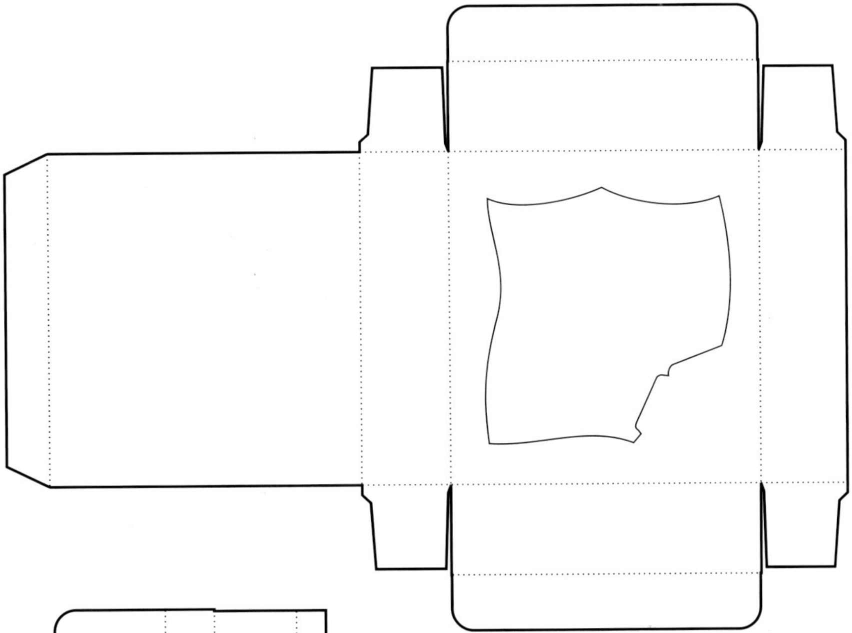

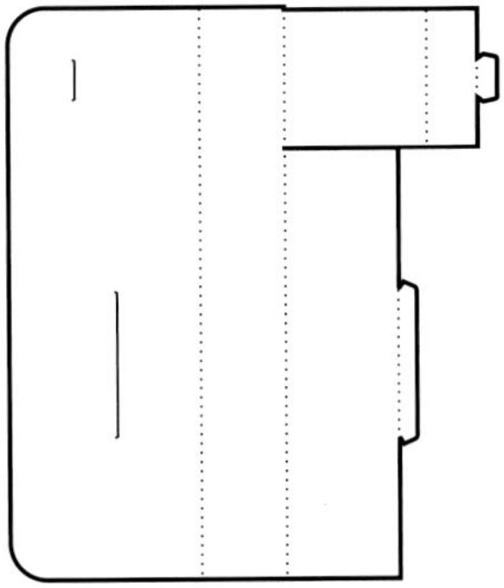

北欧风情
GAOJITANGGUO
情诗什锦糖果

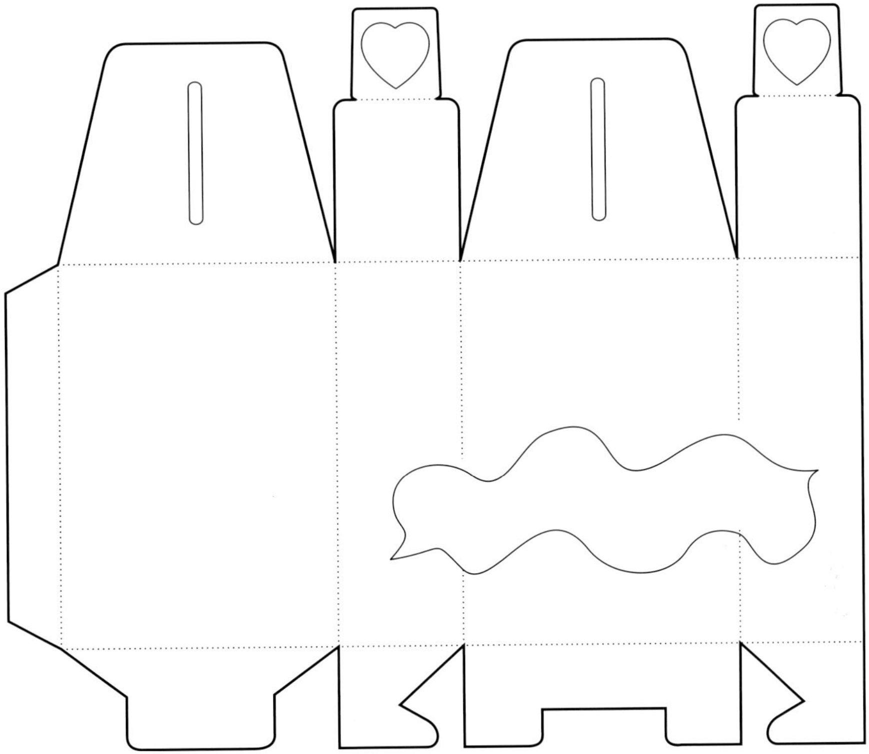

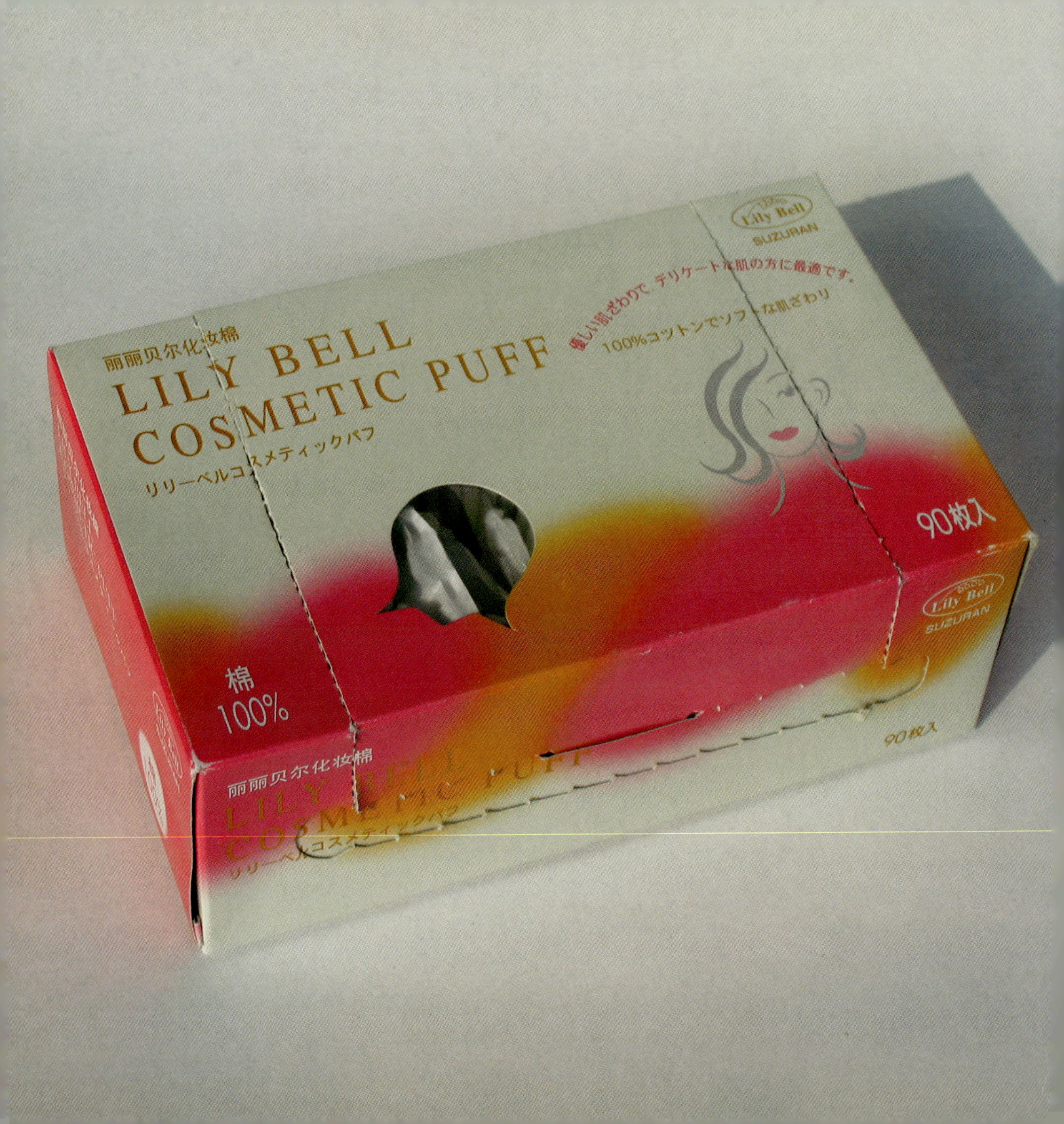

丽丽贝尔化妆棉
LILY BELL
COSMETIC PUFF
リリーベルコスメティックパフ
優しい肌ざわりで、デリケートな肌の方に最適です。
100%コットンでソフトな肌ざわり
Lily Bell
SUZURAN
90枚入
棉
100%

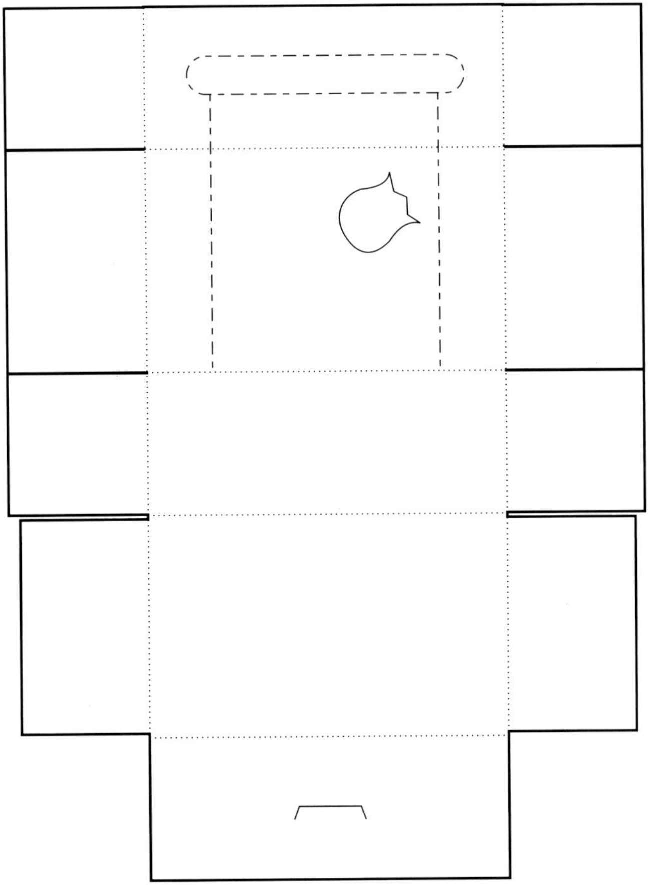

Progresso®
KOH-I-NOOR
12
GRAFITOVÝCH TUŽEK
WOODLESS GRAPHITE STICKS
GRAPHITSTIFTE IM LACK
AYONS GRAPHITE À VERNIS
8911/8B
CE
8911
12

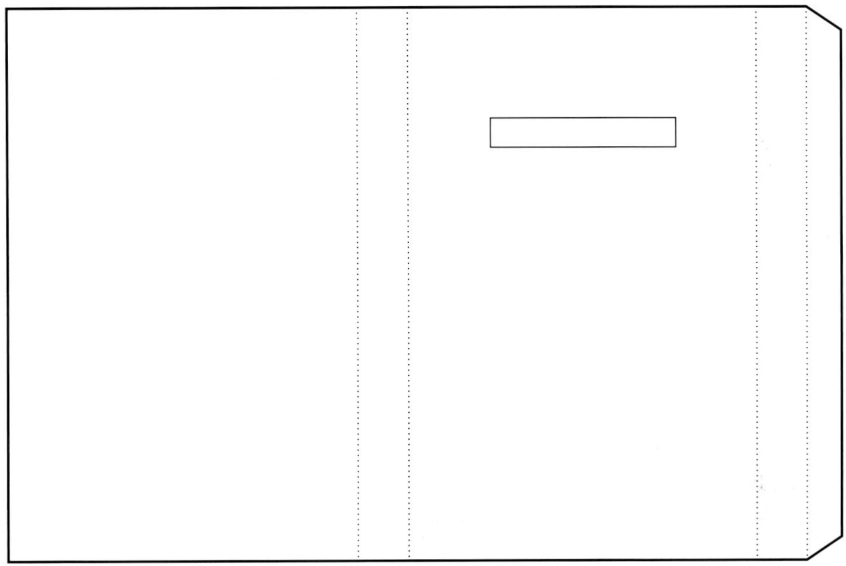

团圆佳节
快乐中秋 相聚团圆
The Mid-Autumn Festival

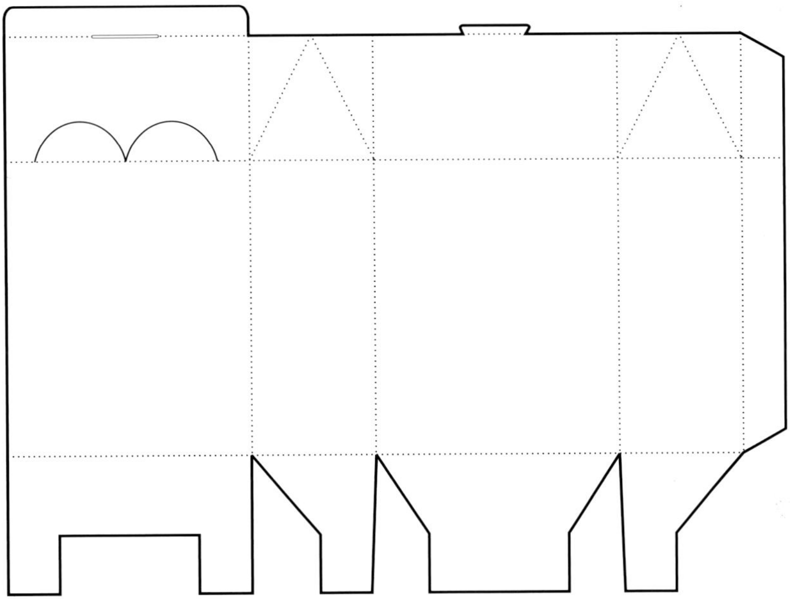

Moon Cake

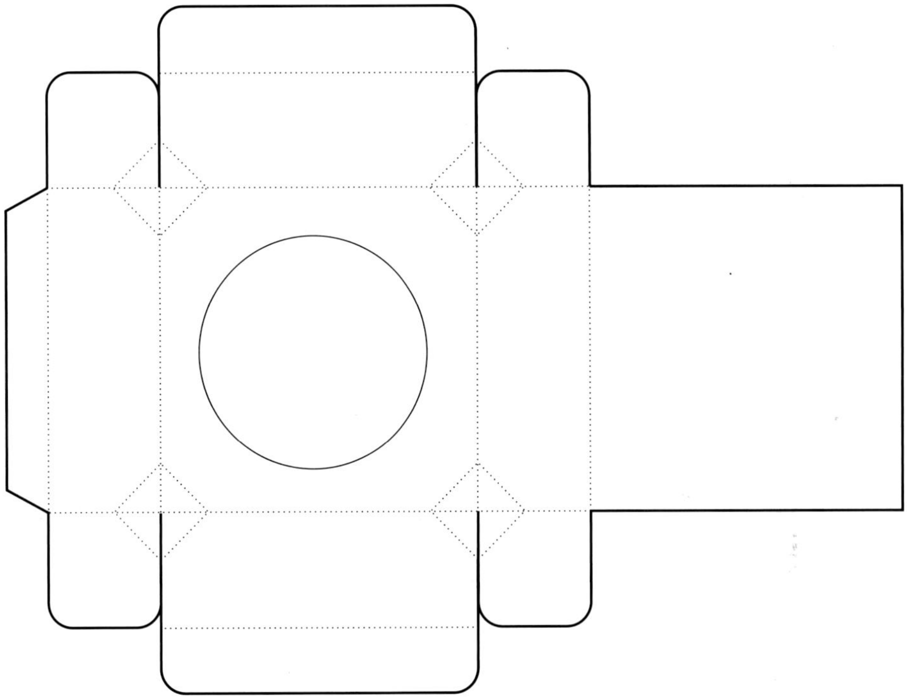

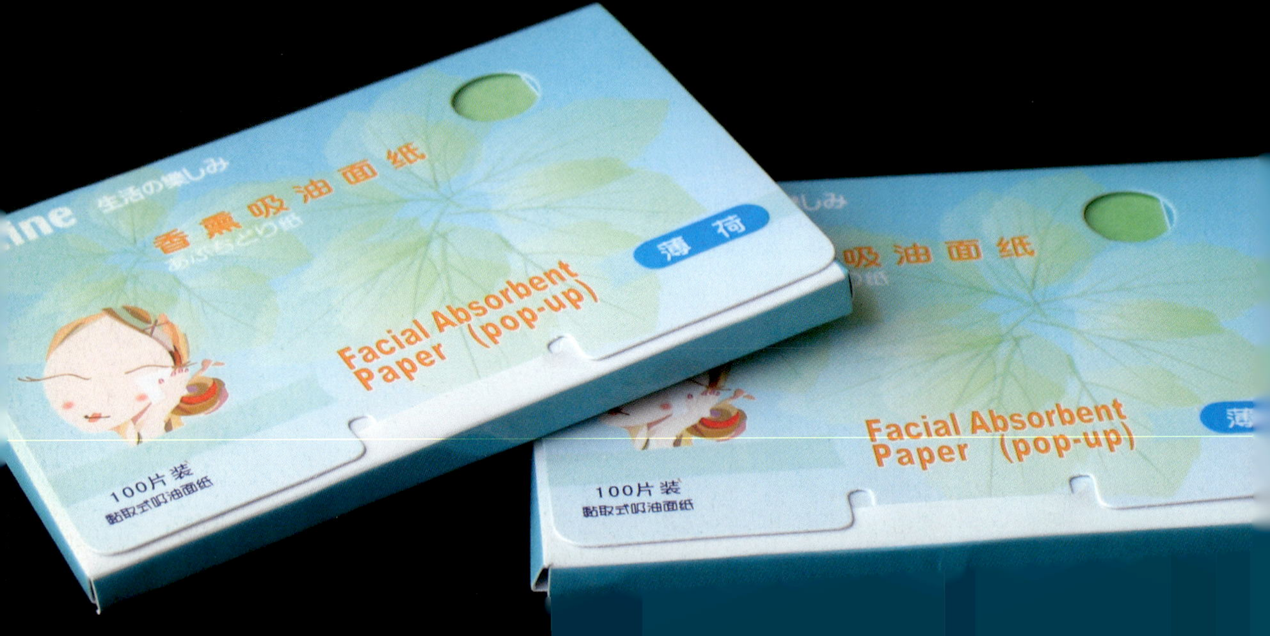
生活の楽しみ
香薰吸油面纸
あぶらとり纸
薄荷
Facial Absorbent
Paper（pop-up)
100片装
黏取式吸油面纸
吸油面纸
Facial Absorbent
Paper（pop-up)
100片装
黏取式吸油面纸

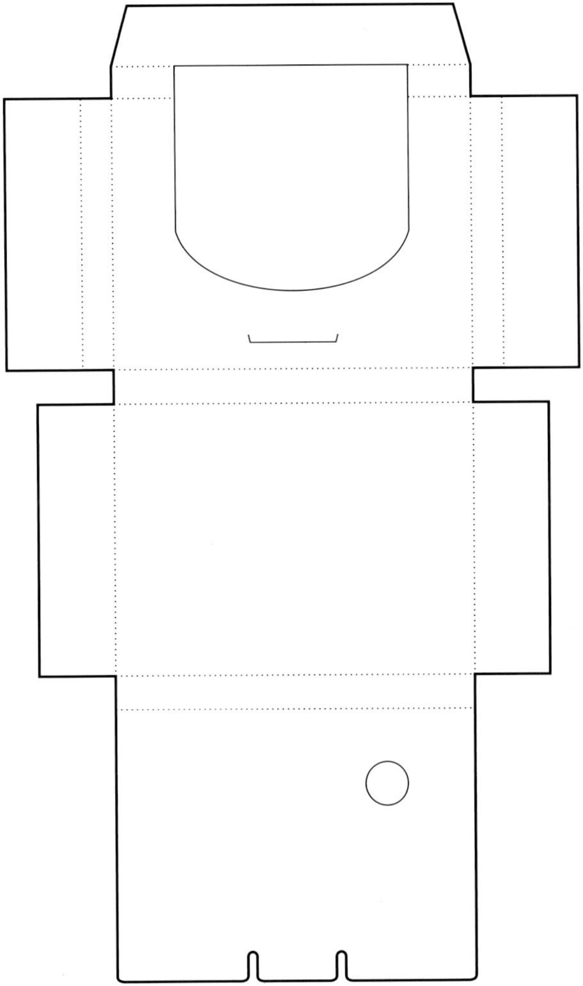

宝富橡皮糖
明治 Meiji
Poifull
葡萄柚味
葡萄味
苹果味
麝香葡萄味
净含量 53克

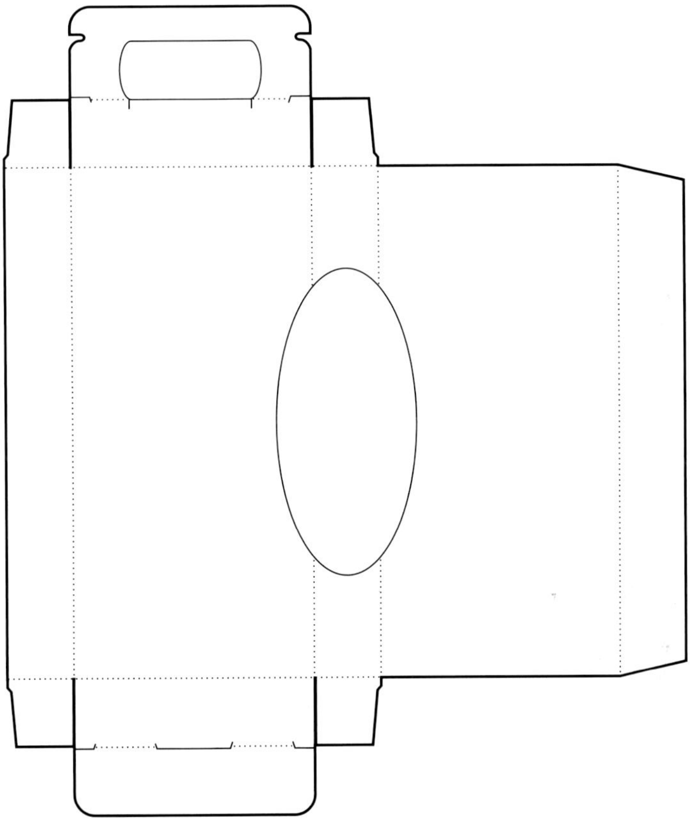

LOTTE
그대만의 향기
이브
XYLITOL
그대만의 향기
한송이
장미처럼....
이브
500

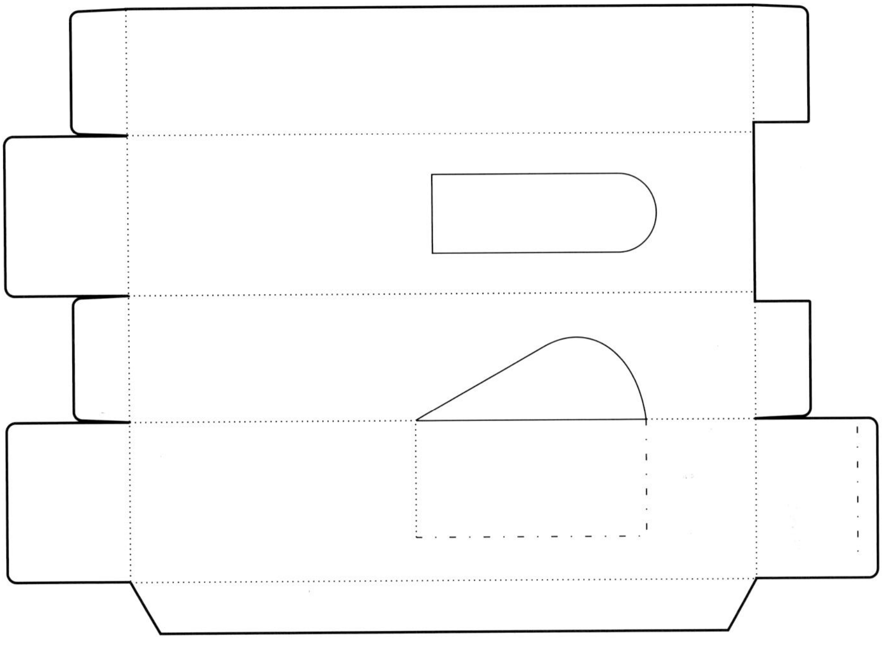

新主张
ISO9001:2000 国际质量体系认证
上海服装行业协会名牌产品
hfä 健一®
女士内衣
2
时尚"媚儿"
莱卡格仔平角裤
WOMEN'S UNDERWEAR
M
94%棉、6%氨纶

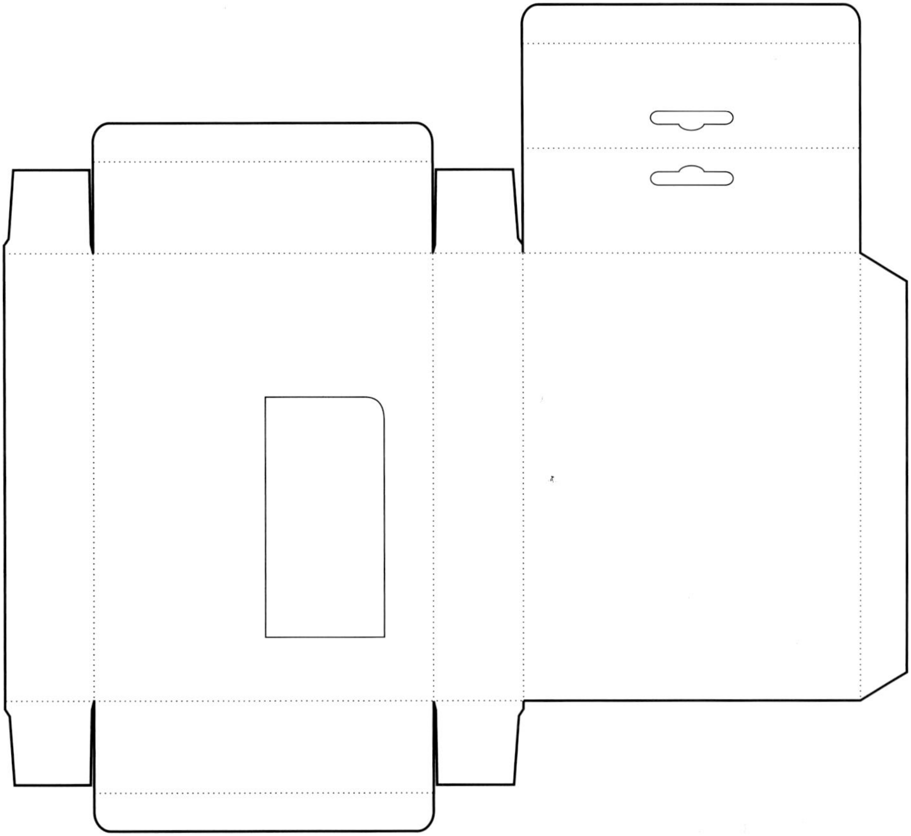

MOST NEW DESIGN
Speed Race
21 CENTURY NEW DESIGN
R/C
SPEED RACER
21 CENTURY NEW DESIGN
New
celerity
MODERN
9V
VOLT
X1
1.5V
AA
X4
REMOTE
CONTROL
QD
FORMAT NOVEL
3+
QD
FORMAT NOVEL
REMOTE
CONTROL
SUPERIOR MATERIAL
SPEED
21 CENTURY

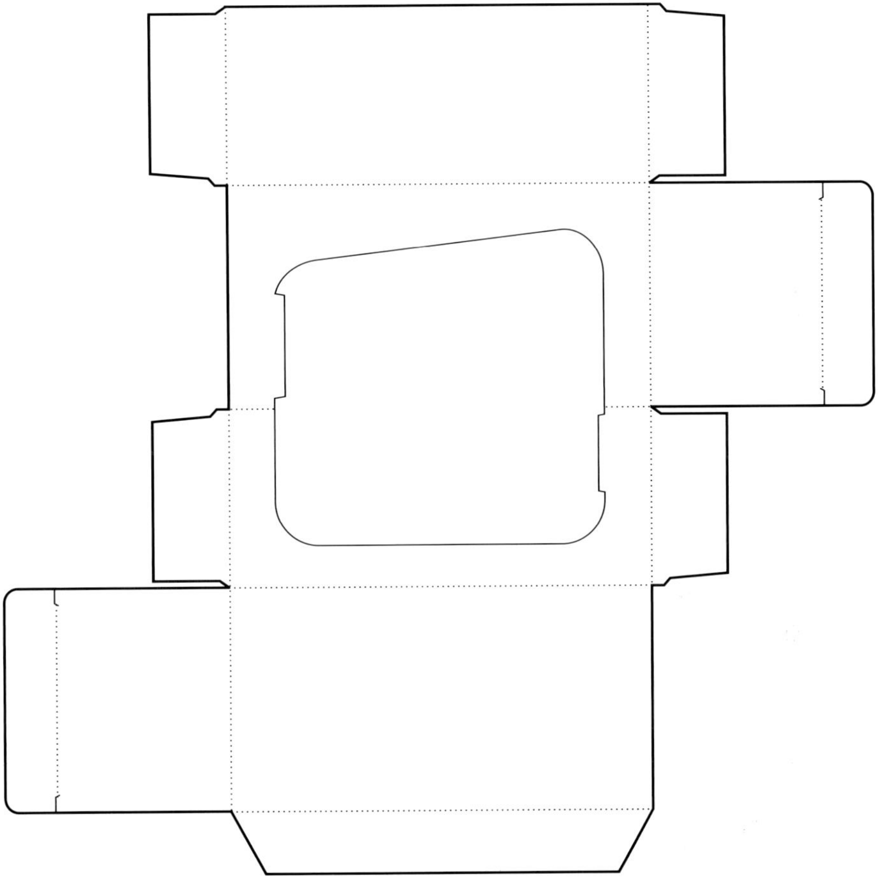

COCO
milk chocolate
酷可 牛奶巧克力
25 块装
packs
Net Wt.: 375g
So Cool!
milk chocolate
酷可 牛奶巧克力
雅客 yake
15g x 25packs

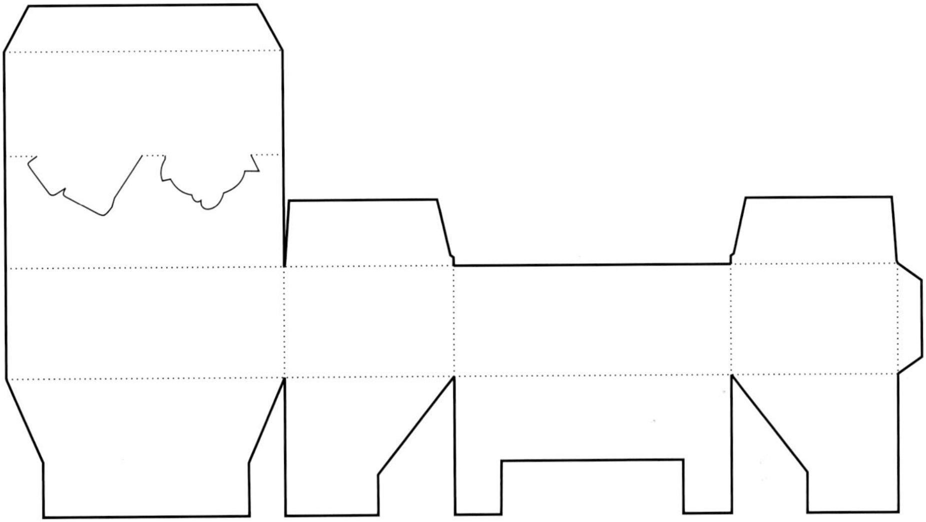

Charger
中国心
China Heart
SCUD
中国心
万能旅行充电器
Omnipotence Travel Charge
SC-537A
·适用于多种手机电池/小灵通电池充电
CHINA HEART

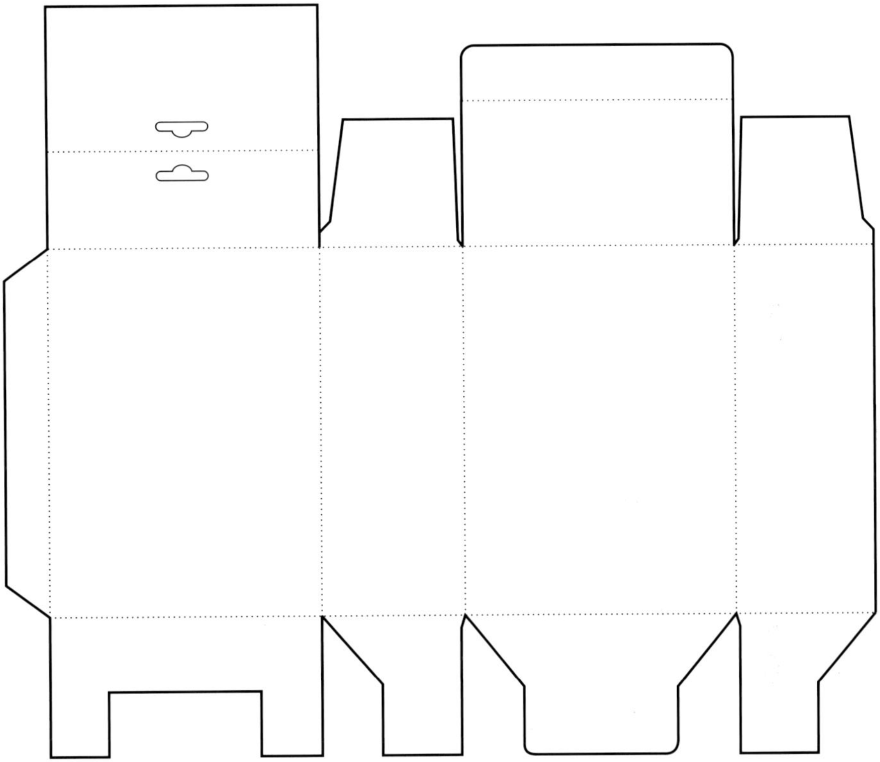

净含量: 12g x 24块
Dove 德芙
由德芙® 牛奶巧克力制成
脆香米®
巧克力
每块只售1元
德芙
牛奶巧克力制成
脆香米®
净含量: 12g x 24块
巧克力
Dove

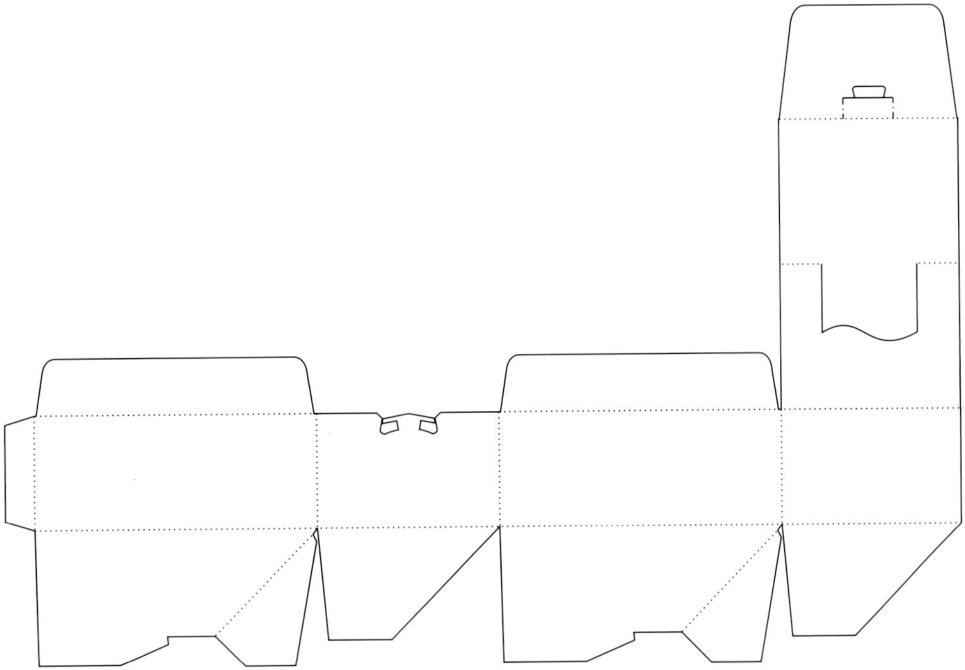

CASIO®
fx-82MS
科学计算
双行显示
统计计算
分数计算
AA型
电池
国家
基础教育
课程标准
实验教材
(北师大版)
(华师大版)
选定
附带插拔式硬盒
CASIO
fx-82MS

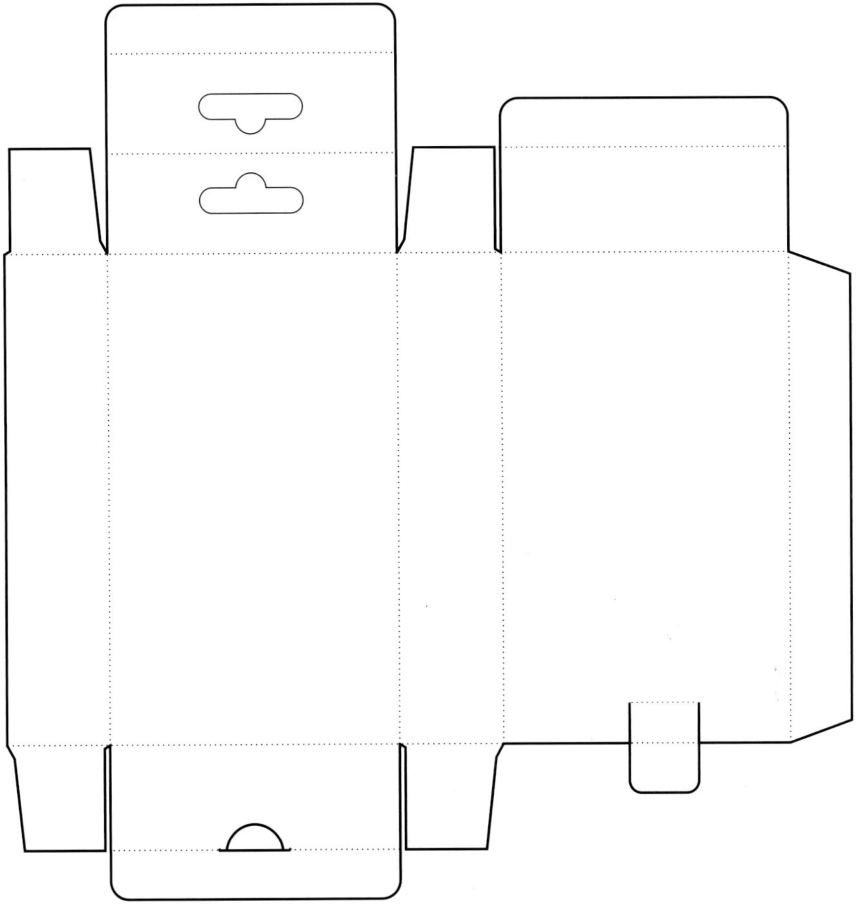

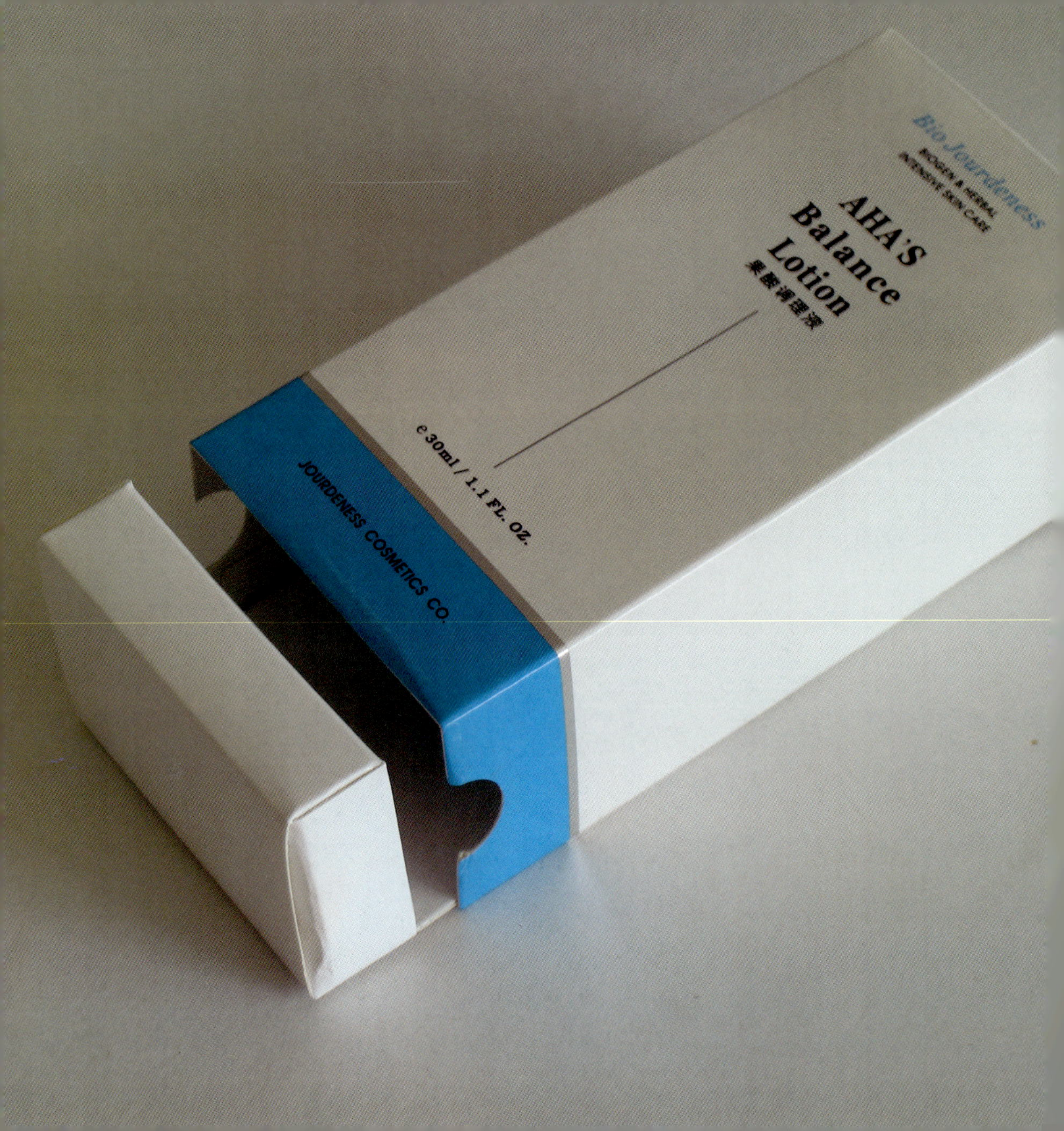
Bio Jourdeness
BIOGEN & HERBAL
INTENSIVE SKIN CARE
AHA'S
Balance
Lotion
果酸调理液
e 30ml / 1.1 FL. OZ.
JOURDENESS COSMETICS CO.

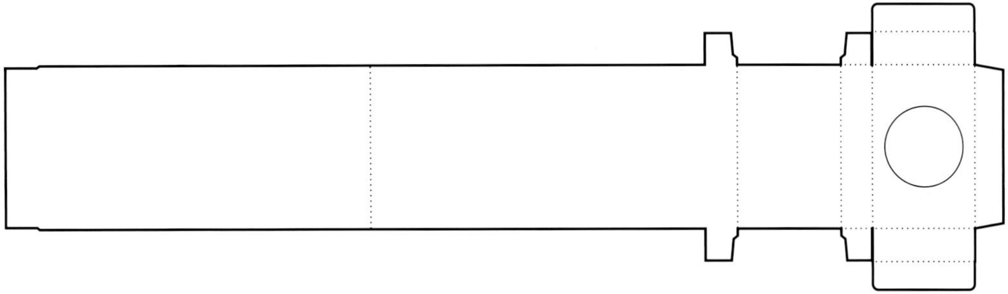

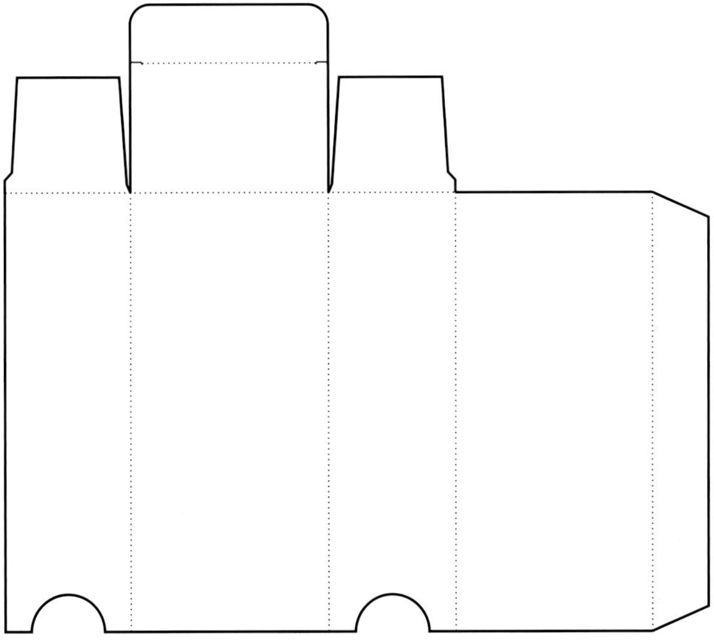

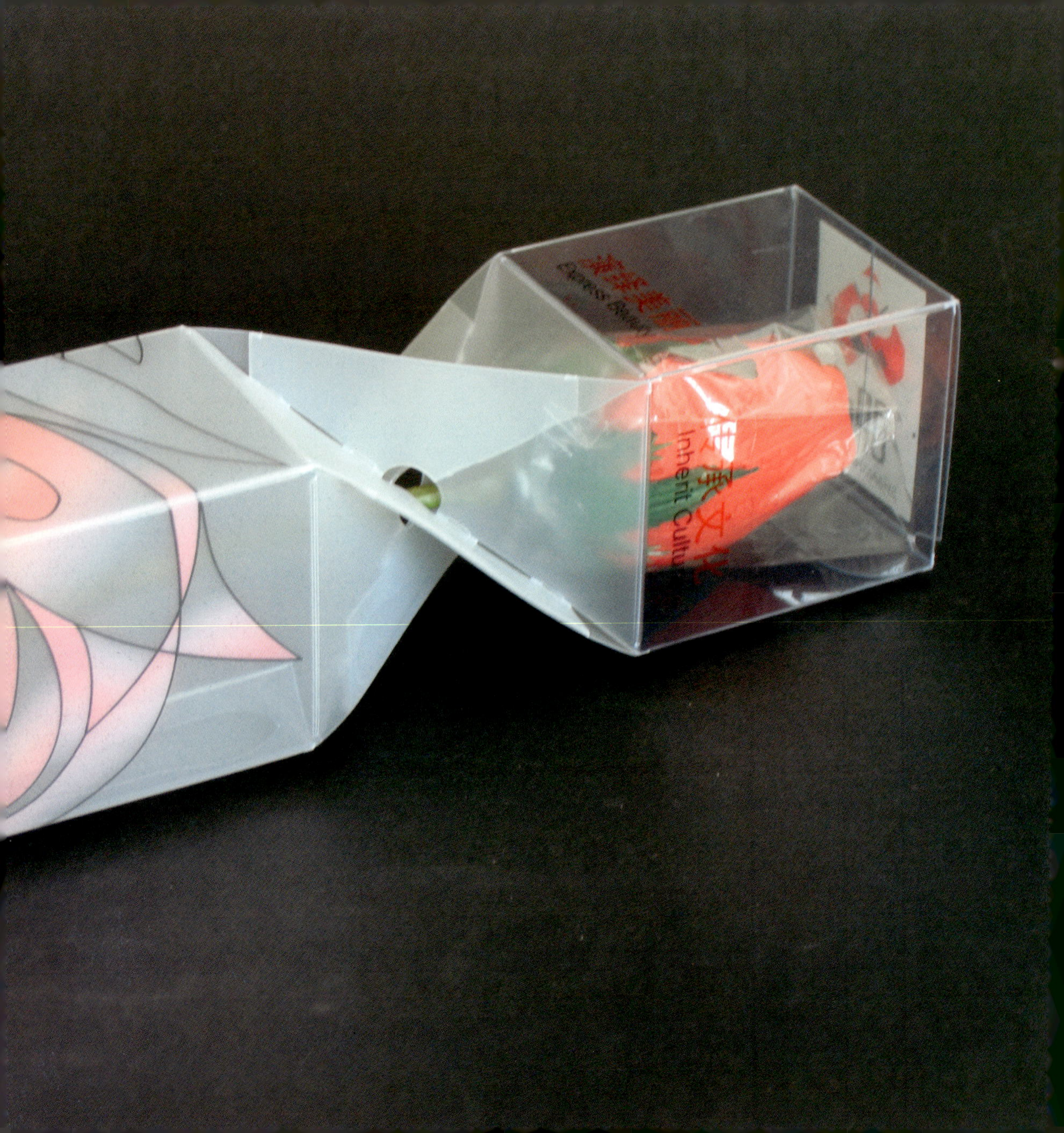
Inherit

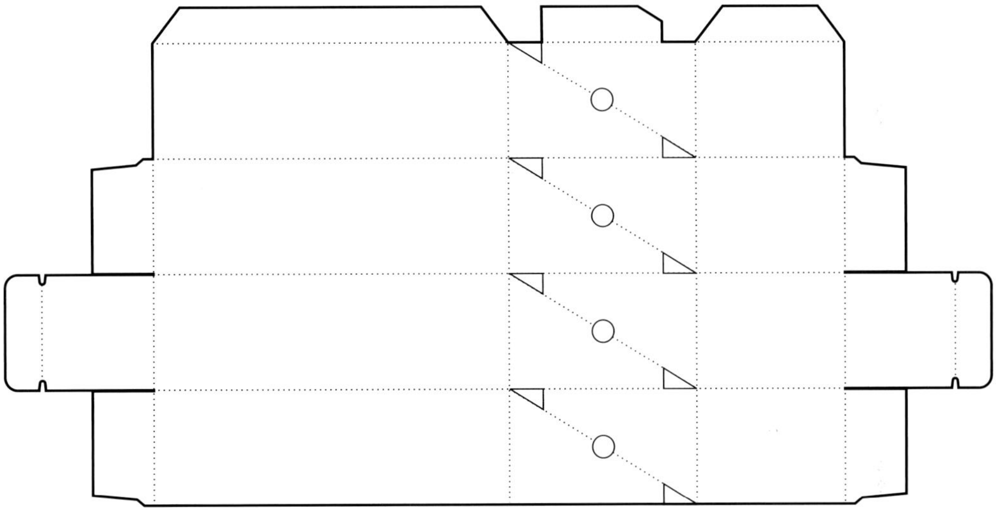

蜂蜜
柚子茶

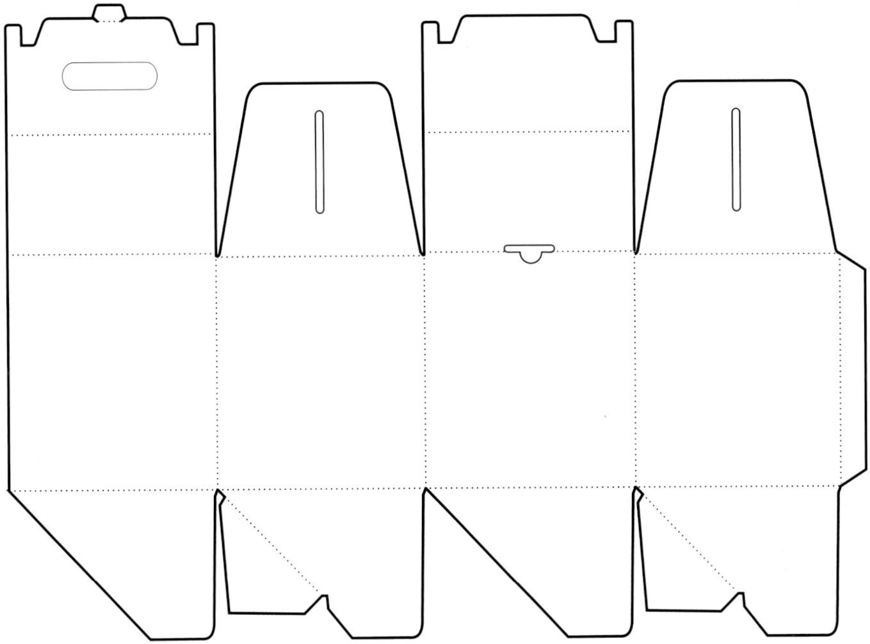

Y3001813

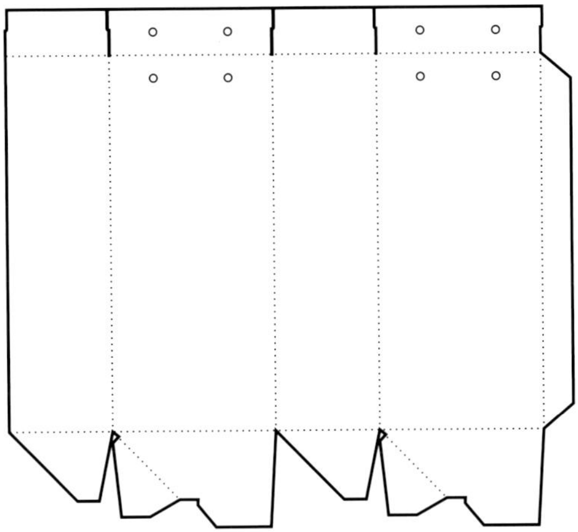

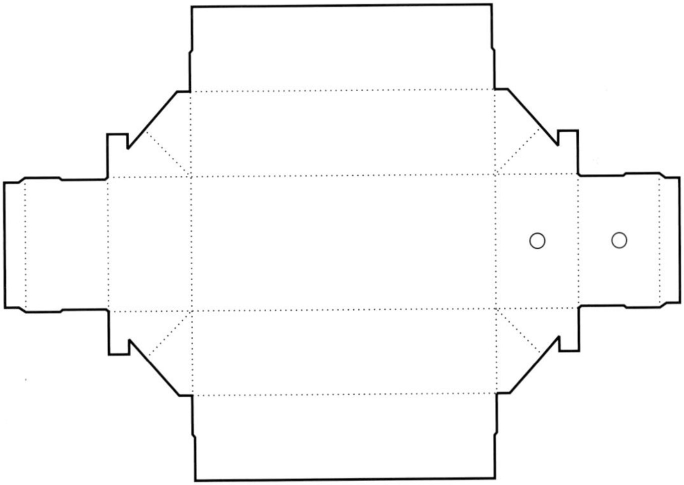

DEFUNG FOOD
上海德丰食品
通过ISO9001国际质量体系认证

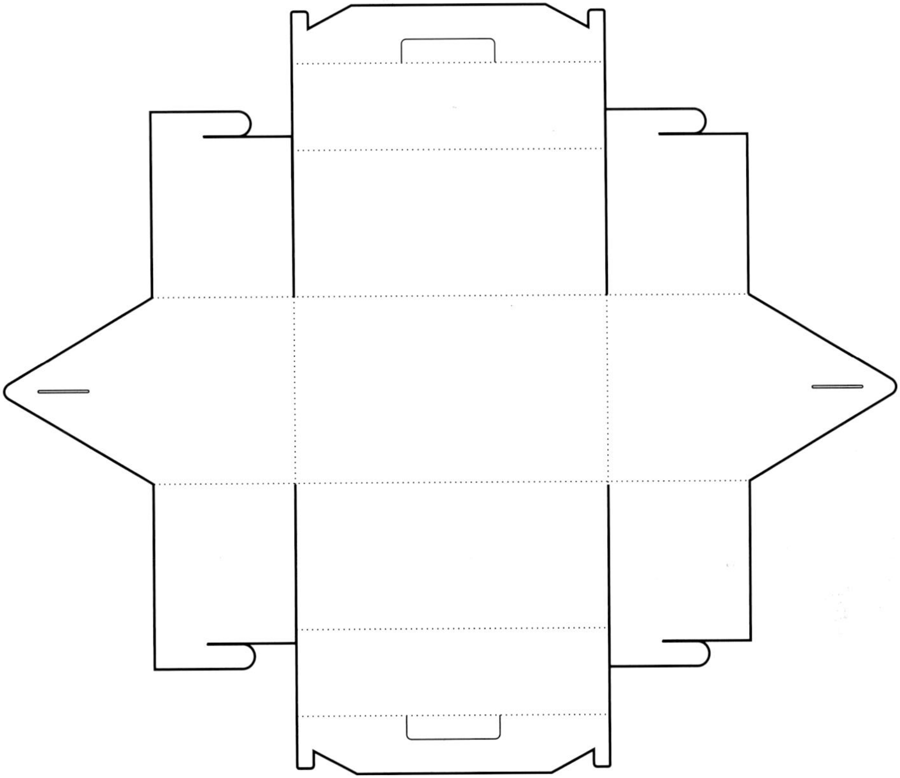

大興八寶粥
TA-HSING
大興
请认准有
陈银清
人像是正宗大兴八宝粥
八寶粥
MIXED CONGEE
傳統食品
360gX12

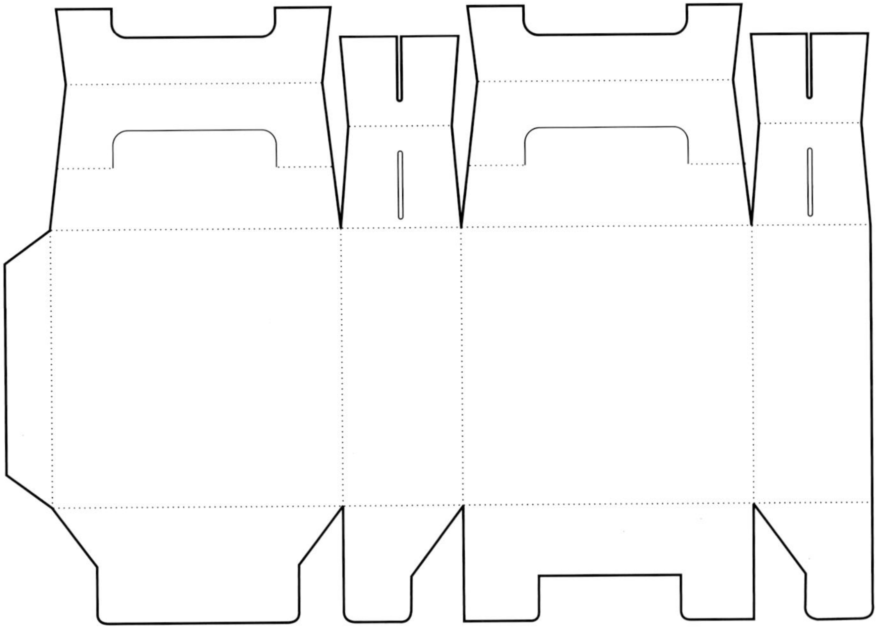

Happy Birthday
蛋糕
吉祥

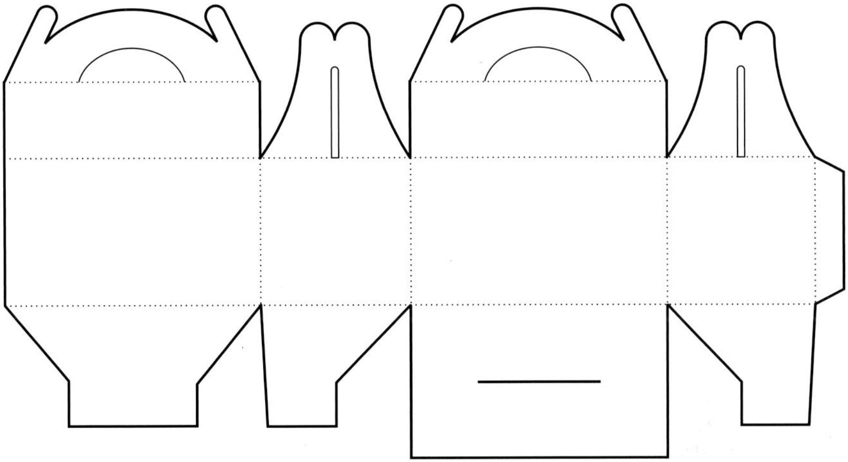

地方特产
中华烘焙明星饼店
當選月餅・全國十佳
放心品牌・知名企業
餅業世家
燕美家樂食品公司

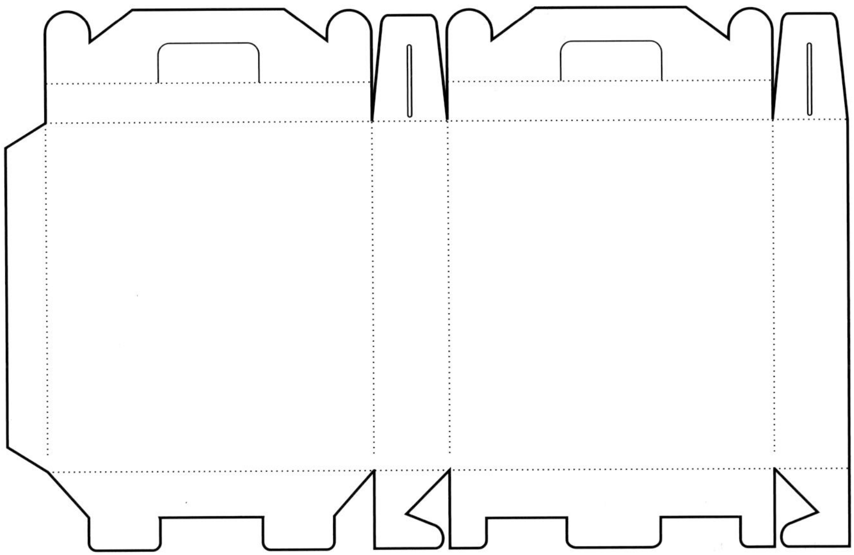

Franklin
洪恩教育集团倾力打造中国儿童“全美语·浸入式”成长环境
洪恩 小乌龟 学美语
适合五岁以上儿童

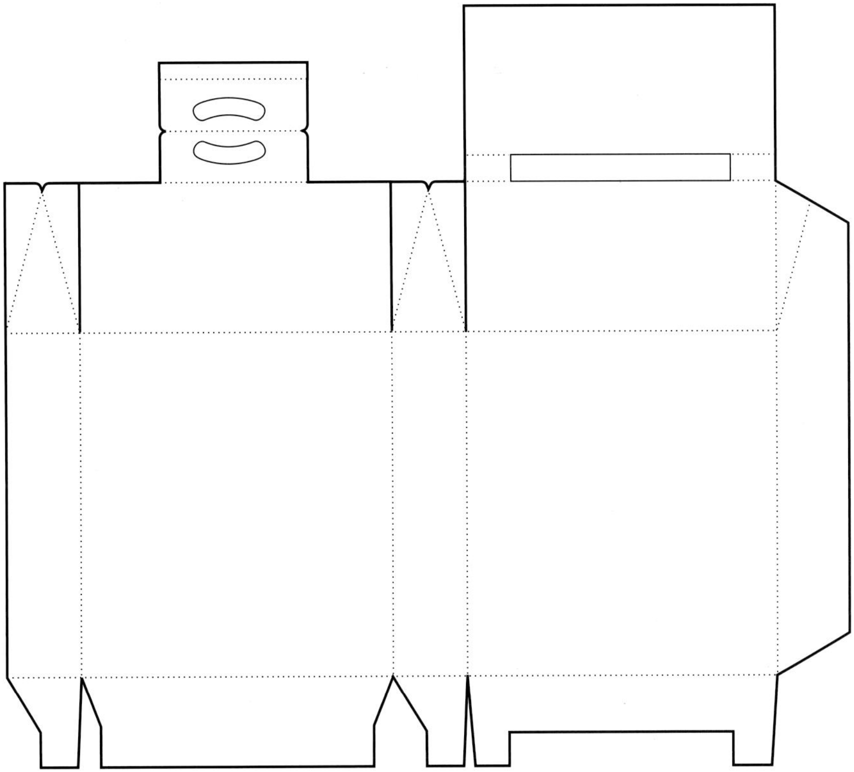

德芙 真藏™ 榛仁夹心巧克力
开启
奥地利进口
国内分包装
净含量:150g

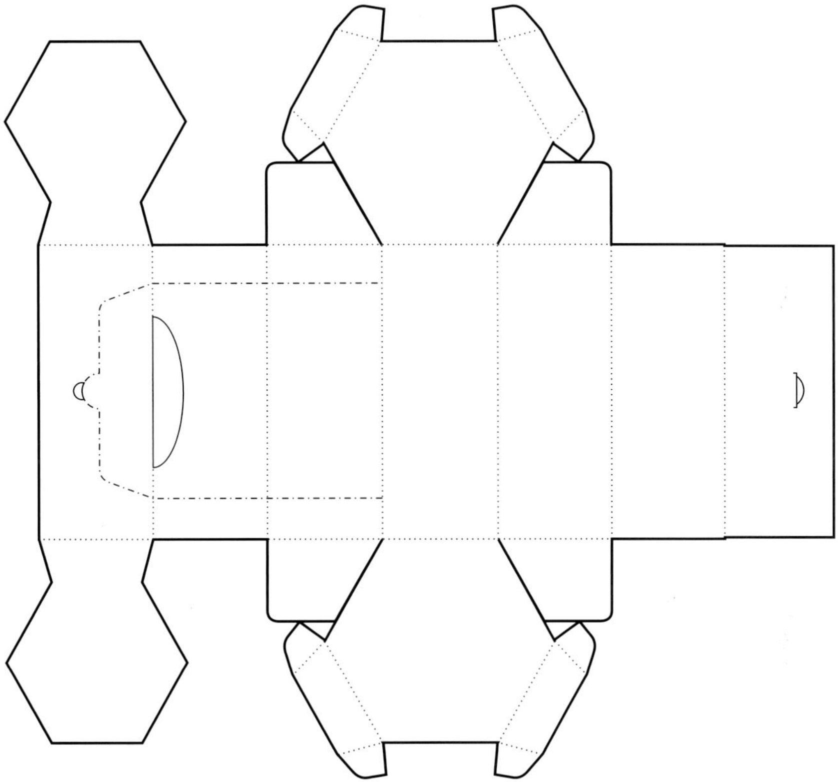

UHA 悠哈®
MONDE SELECTION
BRUSSELS
SELECTION OF QUALITY
珀立妃POLYPHE
纱融牛奶巧克力
POLYPHE珀立妃
纱融系列巧克力获
2006年比利时布
鲁塞尔"世界优质
食品评选会"金奖
净含量100克

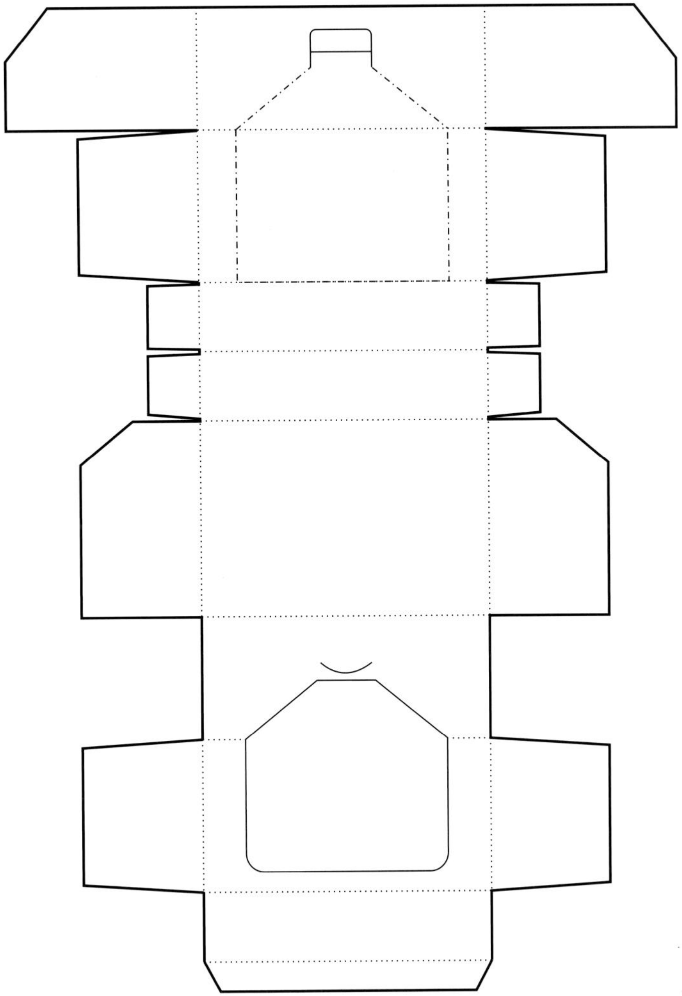

Glico
格力高
"咔力咔力"脆衣杏仁
CALITZ
克粒脆
Almond Chocolate
杏仁巧克力 层层脆型
净含量58g

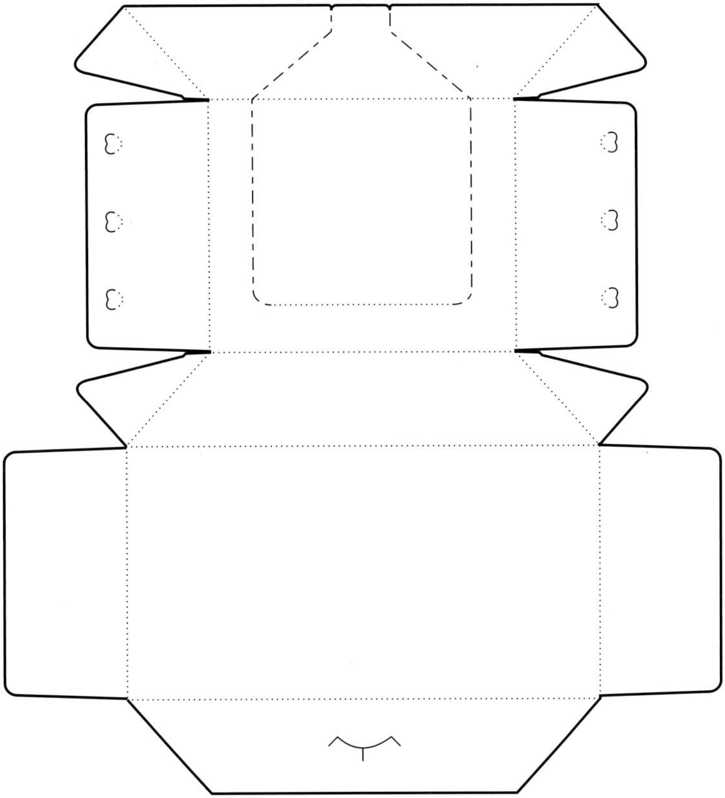

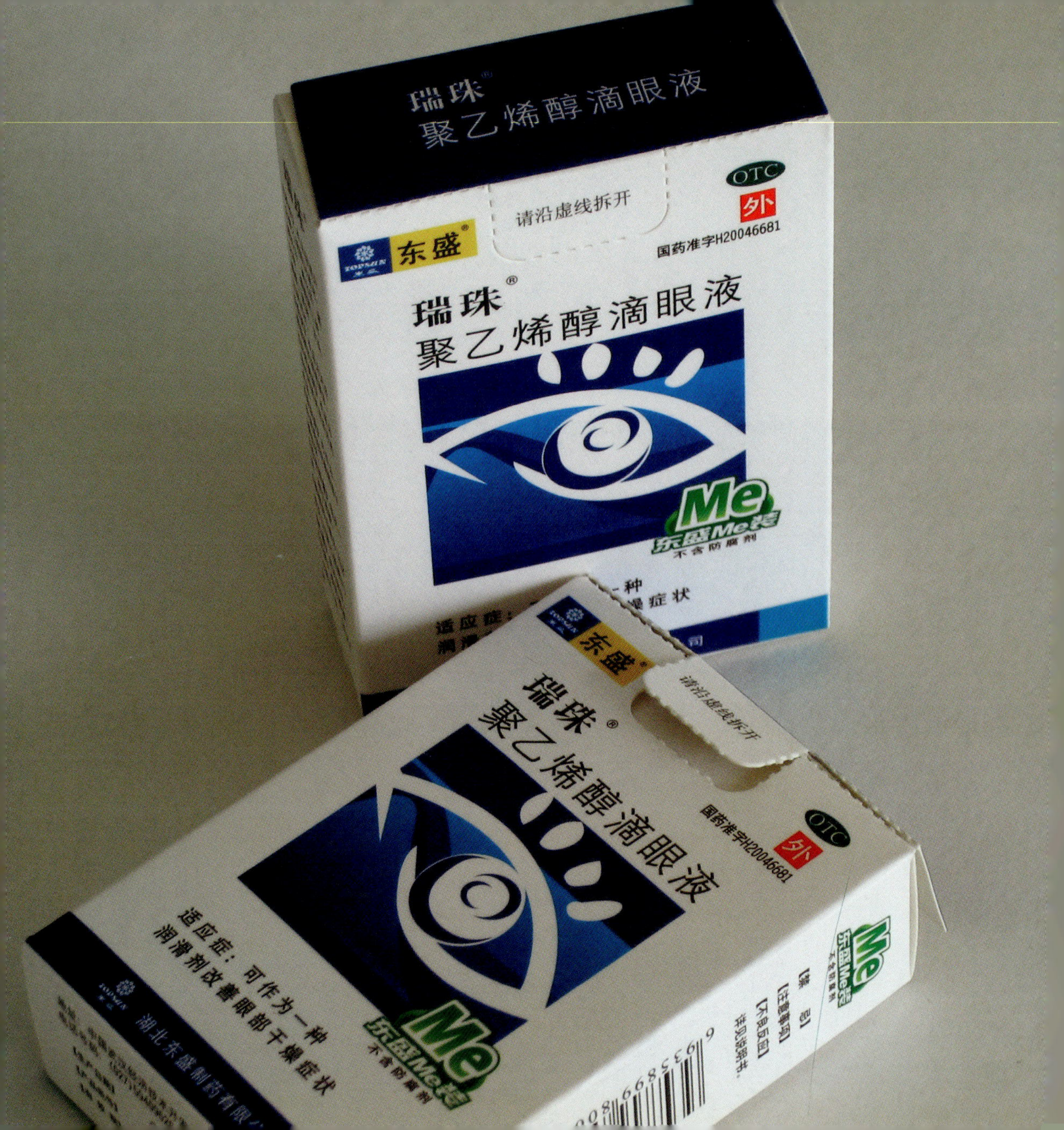

瑞珠®
聚乙烯醇滴眼液
请沿虚线拆开
OTC
外
东盛®
国药准字H20046681
Me
东盛Me装
不含防腐剂
适应症：可作为一种
润滑剂改善眼部干燥症状

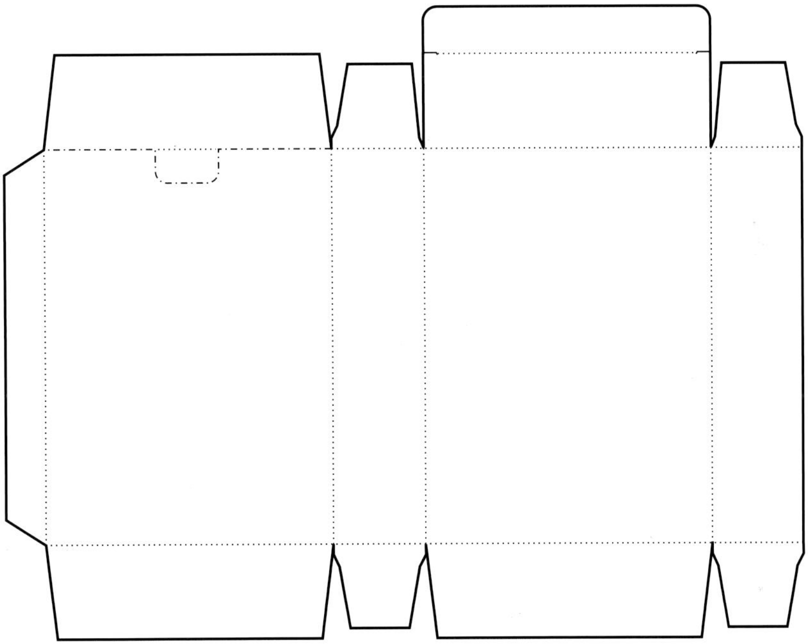

Glico
格力高
pejoy
百醇
巧克力味
夹心饼干
麦芽风味
凝聚成纤巧身形
满心喜悦的巧克力酱
铸就了
醇正百味之百醇
Malt flavor,
slim figure,
and filled with chocolate cream
to your heart's content.
Giving you 100% pure joy and satisfaction.
净含量：48g(两袋分装)

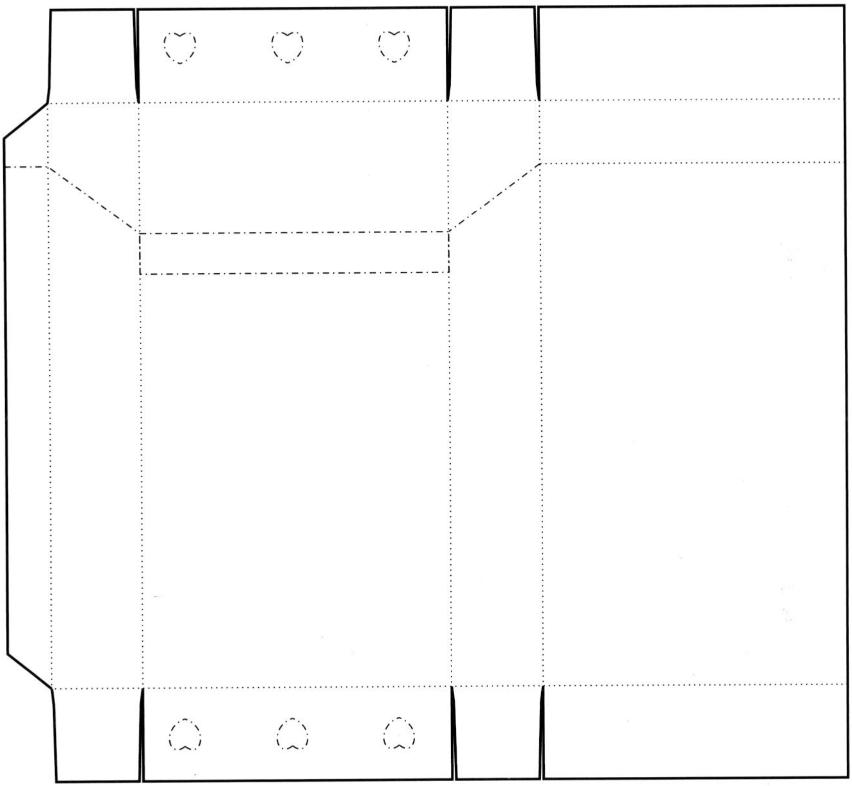

康乃馨 Carnation
高效控水2合1化妆棉
Facial Puff
全新插袋式设计
60+15片装
棉絮不残留
触感更细腻
康乃馨 Carnation
高效控水2合1化妆棉
You Look Different Because You Feel Different
Facial Puff

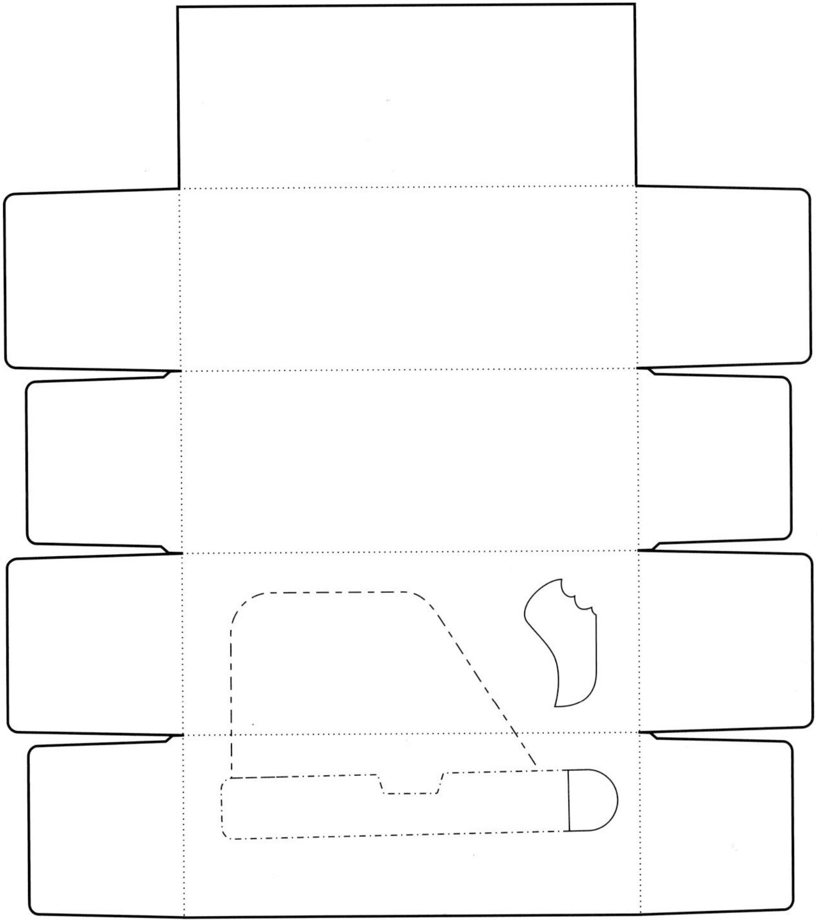

SINKY®
星奇®
星心粒
glico
格力高®
SINKY®
星奇®
星心粒
巧克力味
夹心饼干
净含量: 60g

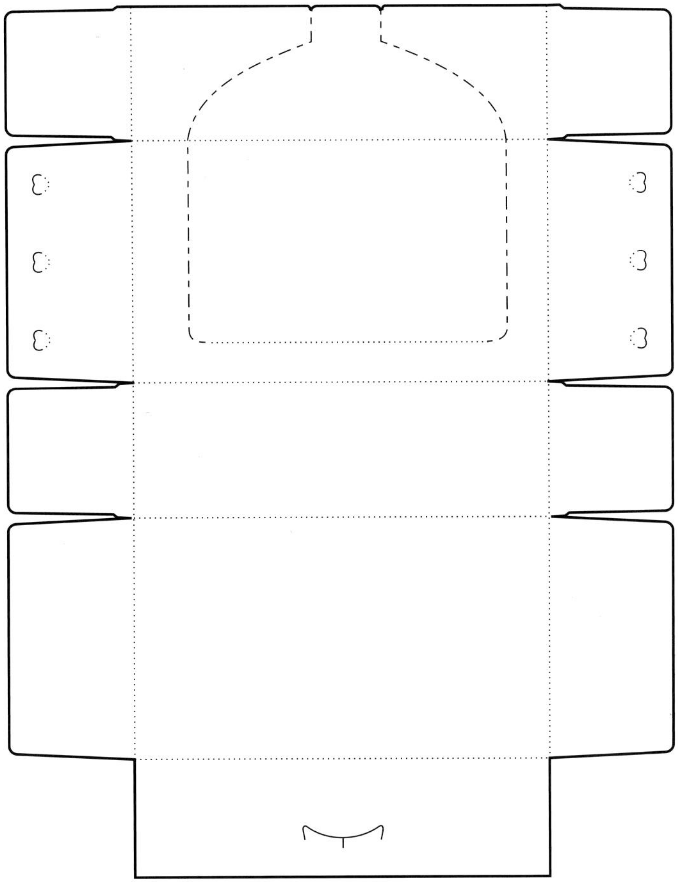

Mousa™
慕纱
GreenTea Chocolate
抹茶巧克力
使用抹茶粉
OPEN ▼
净含量60g
GreenTea
Chocolate
抹茶巧克力
Mousa™
慕纱

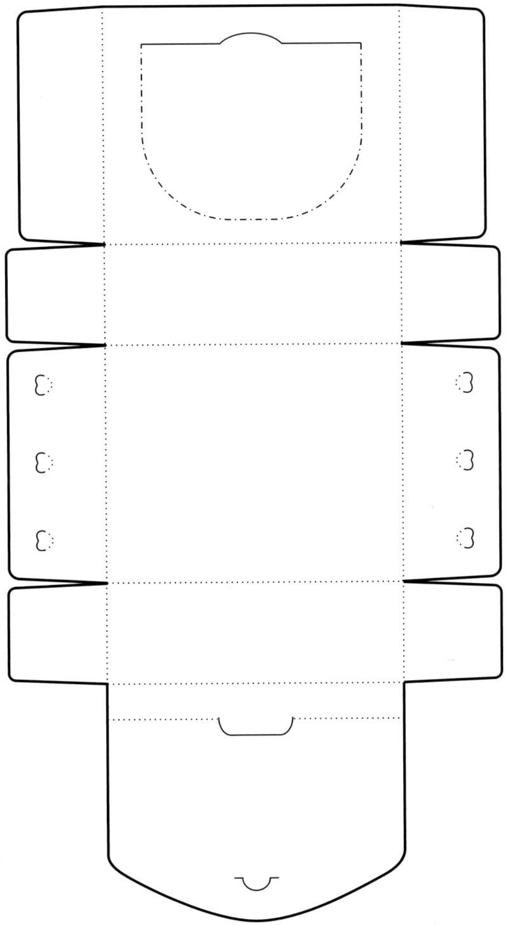

Dove®
德芙® 真藏™
榛仁夹心巧克力
德芙® 真藏™
榛仁夹心巧克力
香浓的德芙®牛奶巧克力，
包裹着酥脆威化卷，珍藏绵密香滑的榛仁夹心。
德芙® 真藏™ 榛仁夹心巧克力
开启
Dove

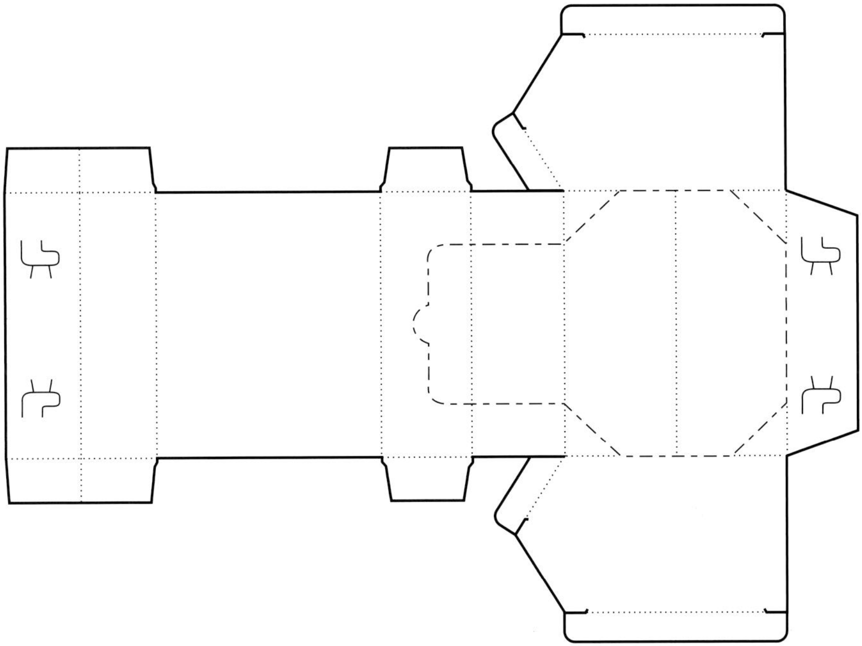

m&m's

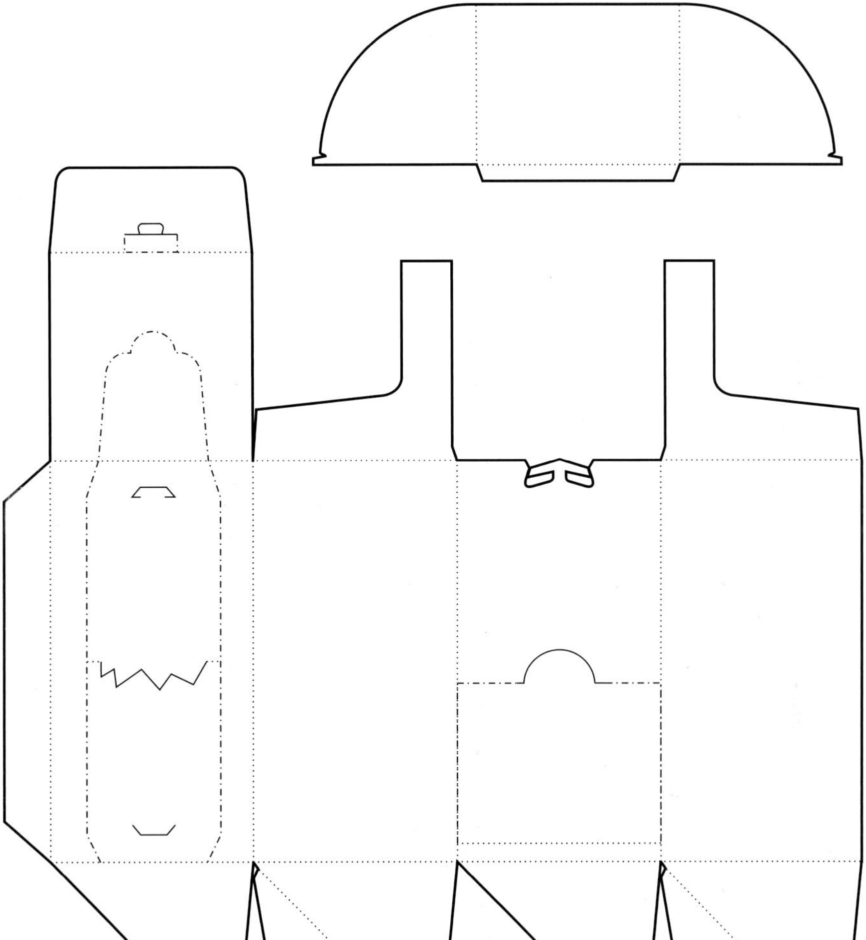

曼妥思
曼妥思
mentos
维加橘园
维C加
脆皮软心球

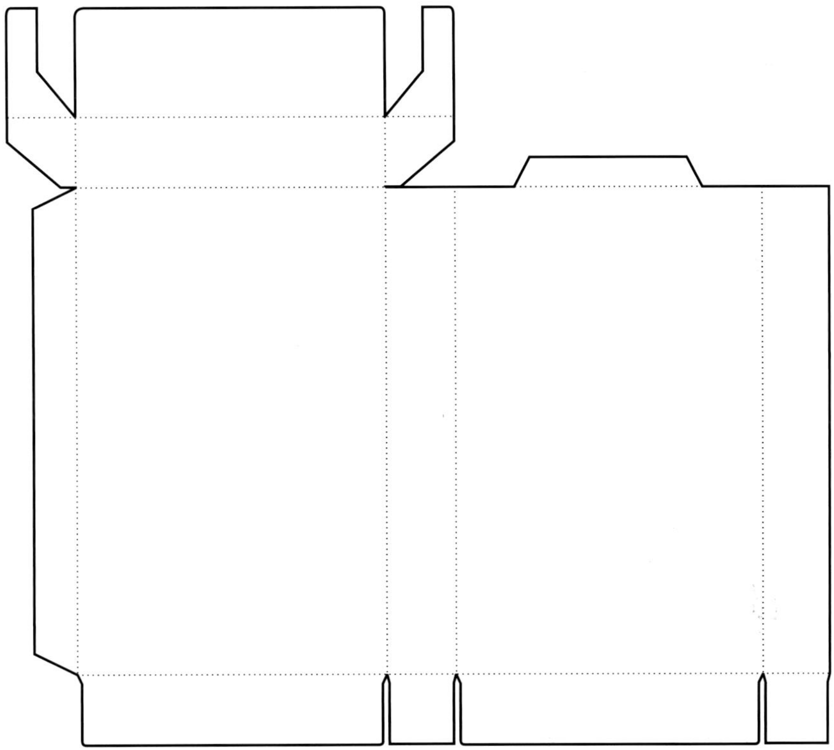

COOL!!
ATTENTION!!
NORDIC SKIES
MICKEY
MICKEY MOUSE
Kleenex
舒洁

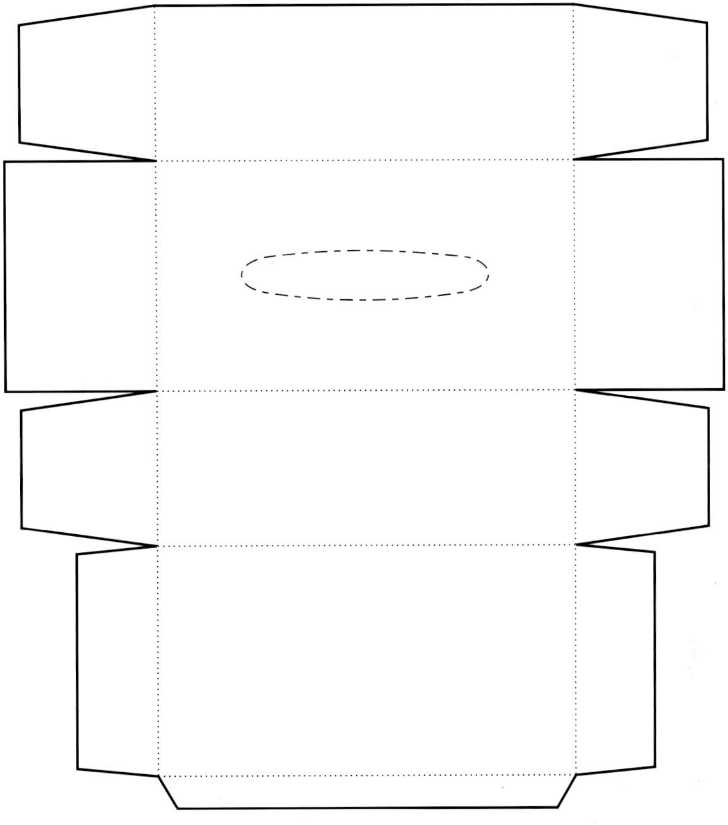

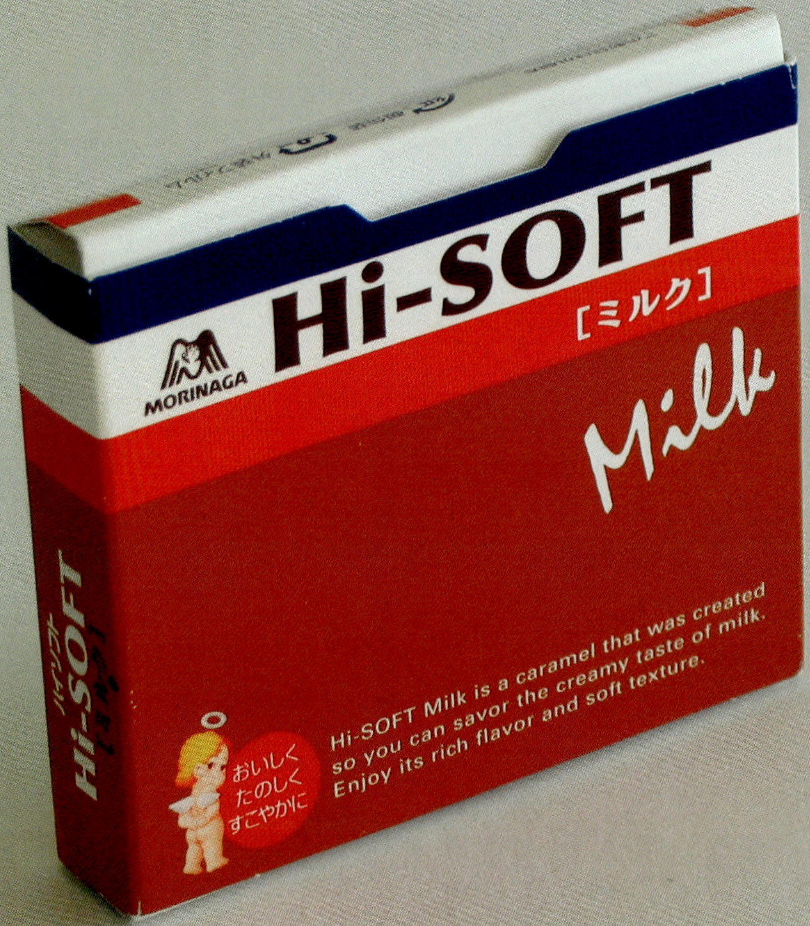
MORINAGA
Hi-SOFT
[ミルク]
Milk
Hi-SOFT Milk is a caramel that was created so you can savor the creamy taste of milk. Enjoy its rich flavor and soft texture.
おいしく たのしく すこやかに
Hi-SOFT

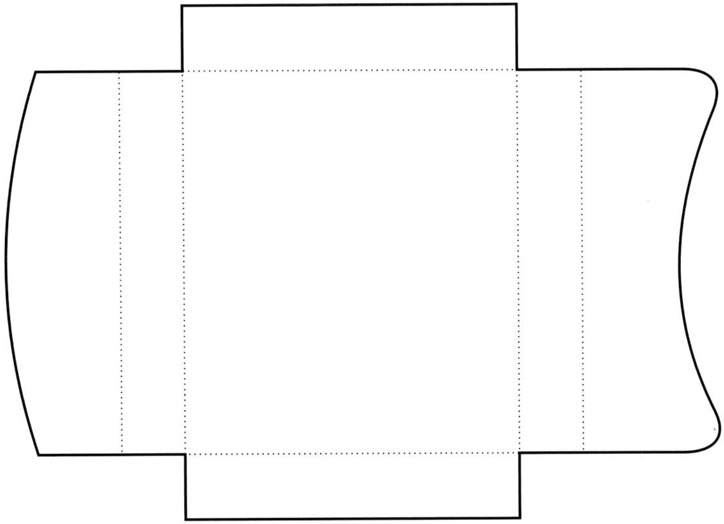

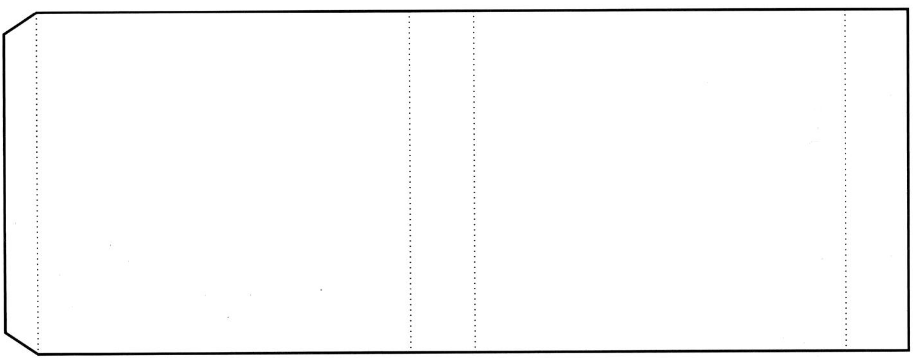

ISETAN
0134-01
¥3000.00
CLEANSE OFF
OIL
HUILE
DEMAQUILLANTE
M·A·C
150 ML / 5.0 US FL OZ ℮

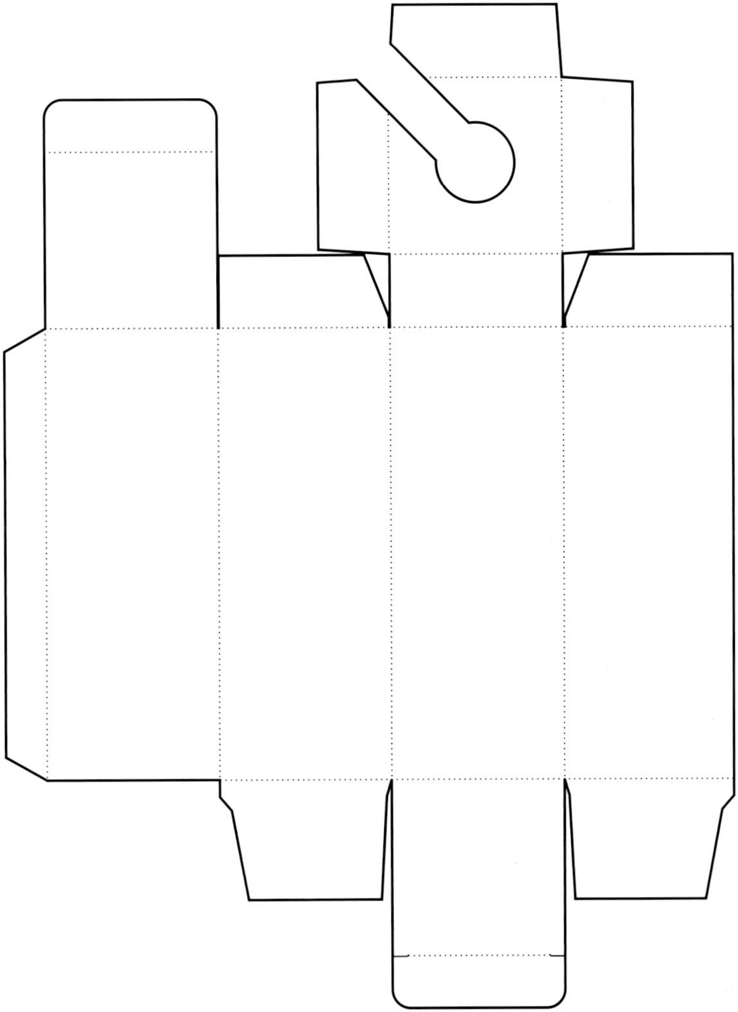

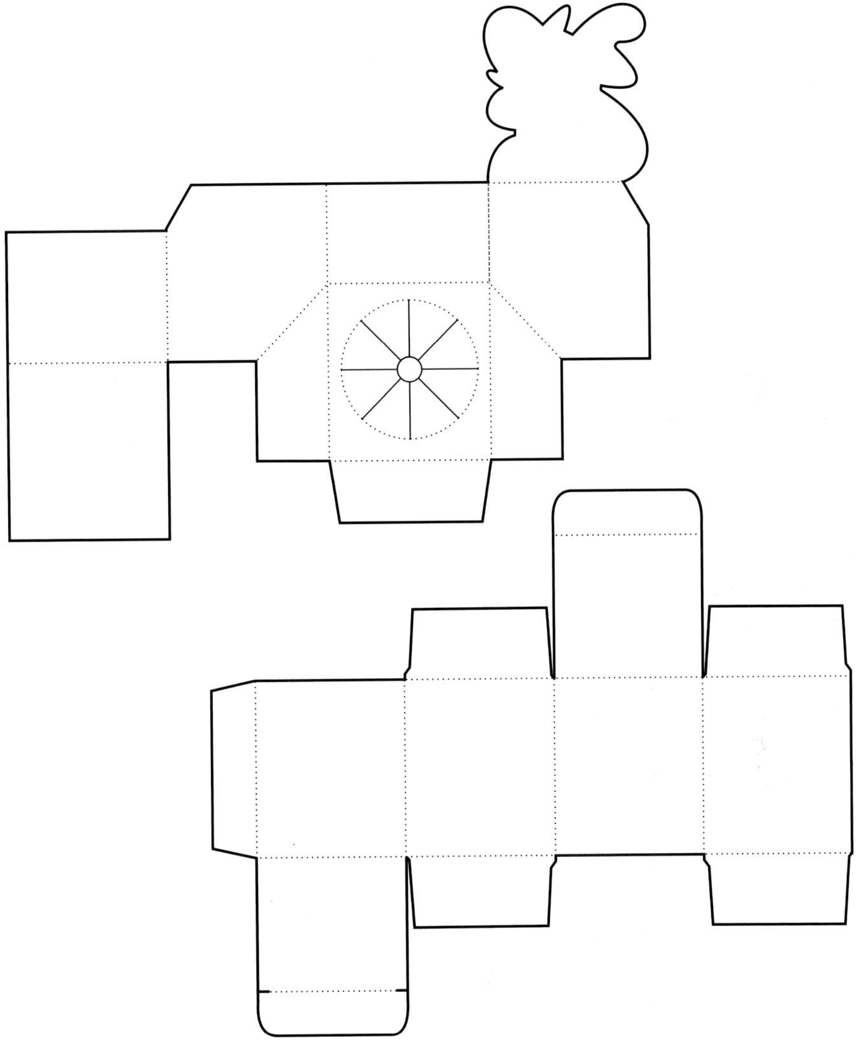

新上市
SINOWAY HERB
相宜本草
安全 有效
深层洁颜油
DEEP CLEANSING OIL
◎蕴含橄榄油／绿茶等本草精华
卸妆去黄 清透滋润
上海中医药大学基础医学院监制
新上市
新上市
修颜乳

www.vichy.com
Melanin, the pigment that darkens
complexion, is produced deep in
the skin by melanocytes and then
loaded in skin cells.
With time and sun exposure, this
melanin load of skin cells
increases: overall complexion
becomes more pigmented and
visibly darker. Brown spots appear
on cheekbones and temples.
Bi-WHITE ADVANCED
The 1st Intensive Whitening
Corrective Care that associa-
tes Procystein, Extra-active
Vitamin C and Liquorice
extract to reduce the melanin load
of the skin and correct darkened
complexion and brown spots.
• Procystein, 10 years of
research, patented molecule,
acts on the chain of melanin
synthesis to reduce its production
by up to 24%*
• Extra-active Vitamin C is a
new extra-active form of stabi-
lized pure Vitamin C. It evacuates
dead melanin filled cells more
rapidly through its cell renewal
THE VICHY SAFETY
COMMITMENT

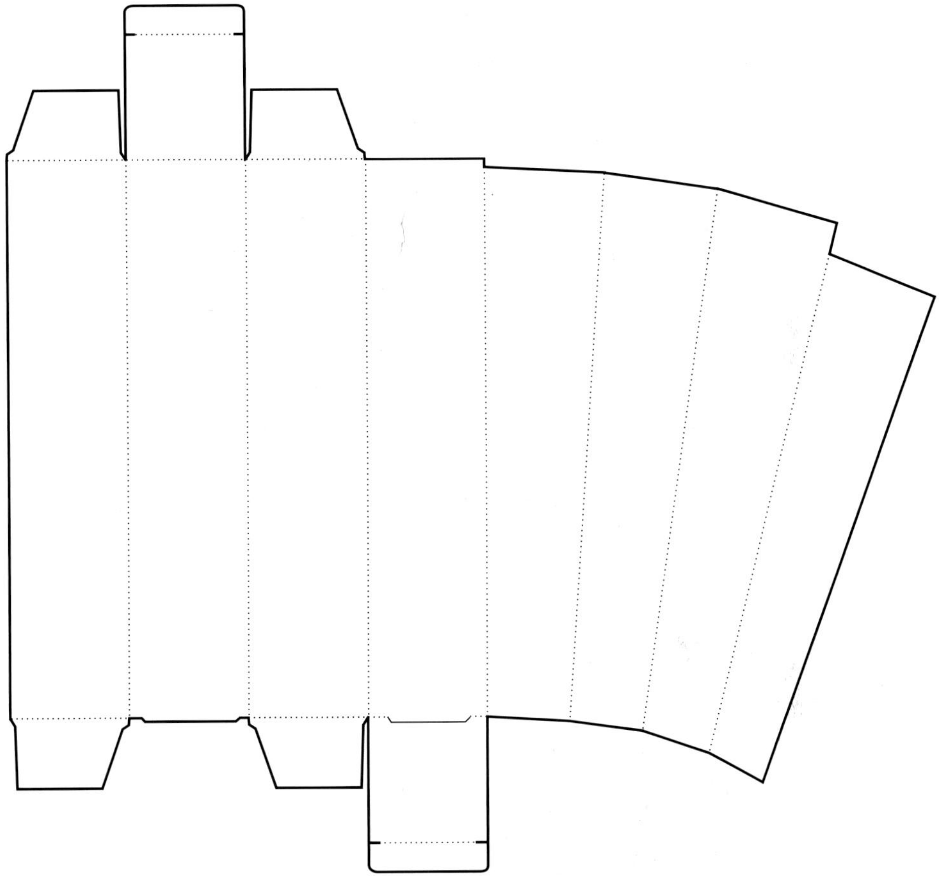

CHEAPANDCHIC
MOSCHINO

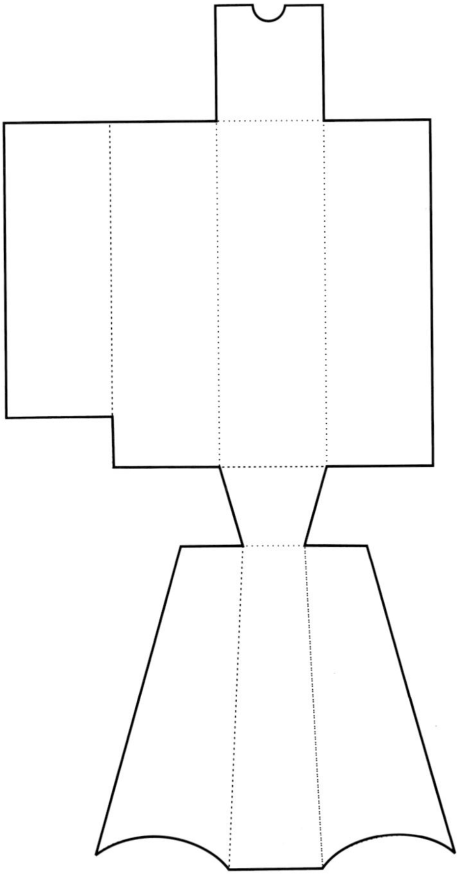

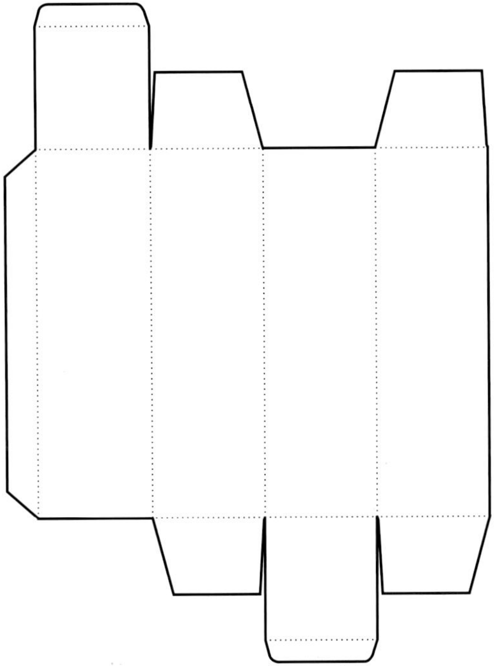

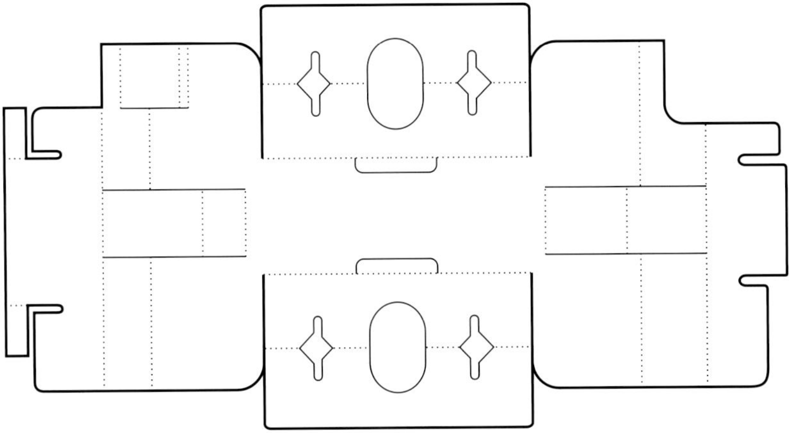

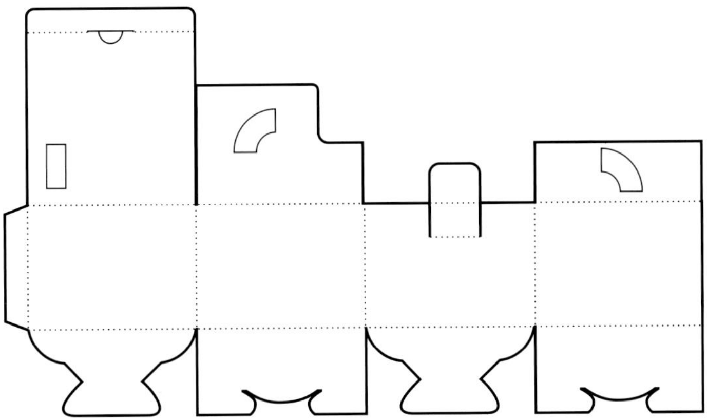

Bio Jourdeness
Foudation Micro-Beads
The biochemical skincare products of
Bio Jourdeness
are developed and introduced
into market by the most advanced
laboratories in the world,such as
France,Switzeri and,Germany,
New York and Japan.
Through a high-tech technique,
the natural essence derived
from animal and plant is
completely condensed and stored.
Its gentle feature tiny molecule,
permeability and strong combination
can efficiently achieve skincare effect
JOURDENESS COSMETICS
&
PERFUME CO.,LTD.

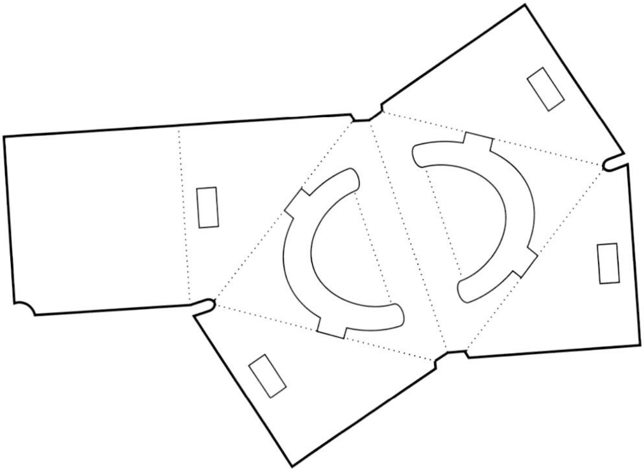

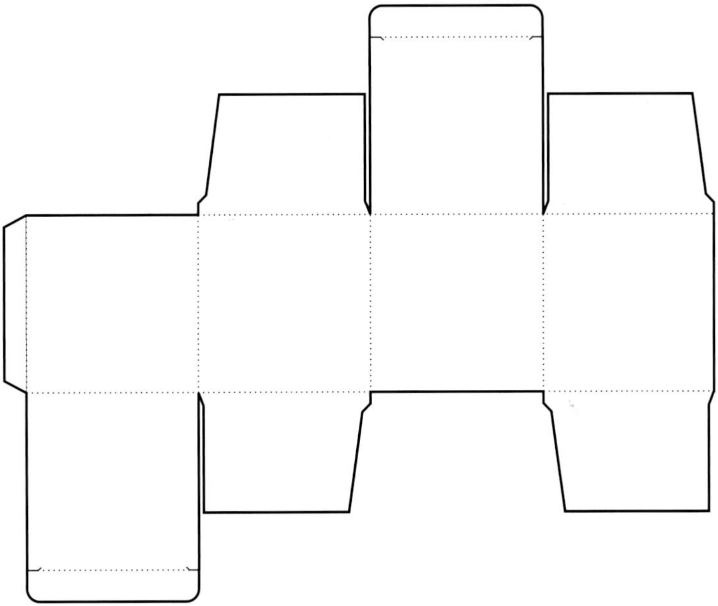

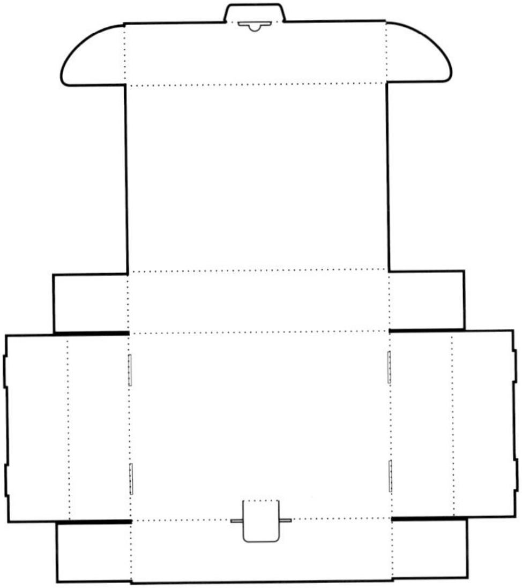

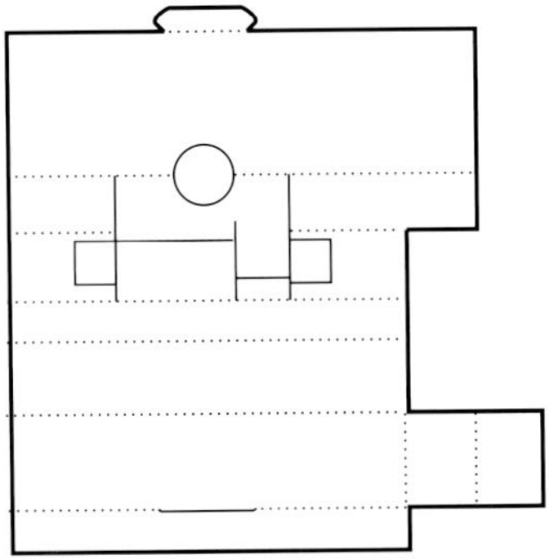

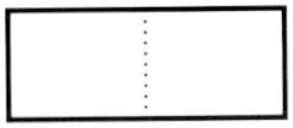

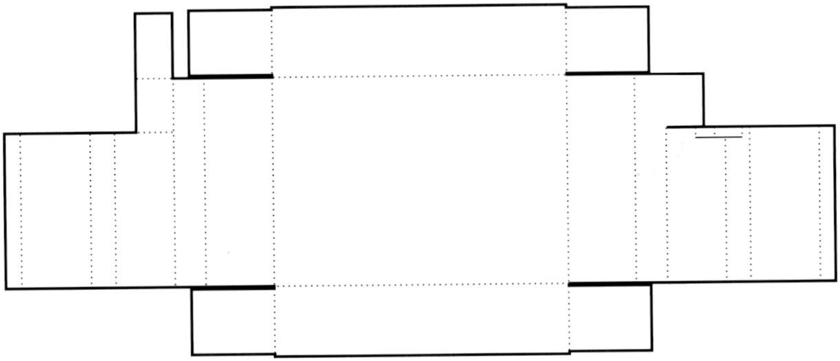

LANCÔME
PARIS
www.lancome.com

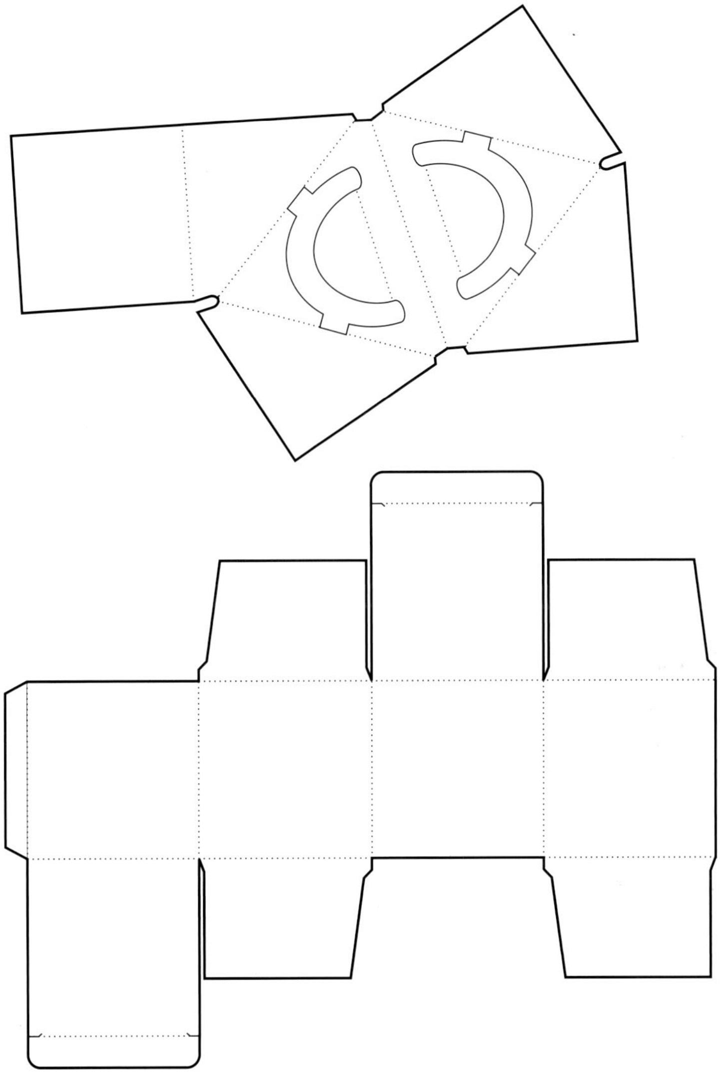

NEC
MEMORY KEY
NU-188
都市精灵
符合USB1.1标准，支持即插即用，热插拔
无需驱动程序（Windows 98除外）
支持BIOS USB-ZIP/HDD引导电脑启动
分区加密功能，带有写保护开关
LED灯指示工作状态
抗震防潮，防电磁波
可擦写1000000次 数据保存达10年以上
无需驱动程序
东电科技
EAST-NEC TECH

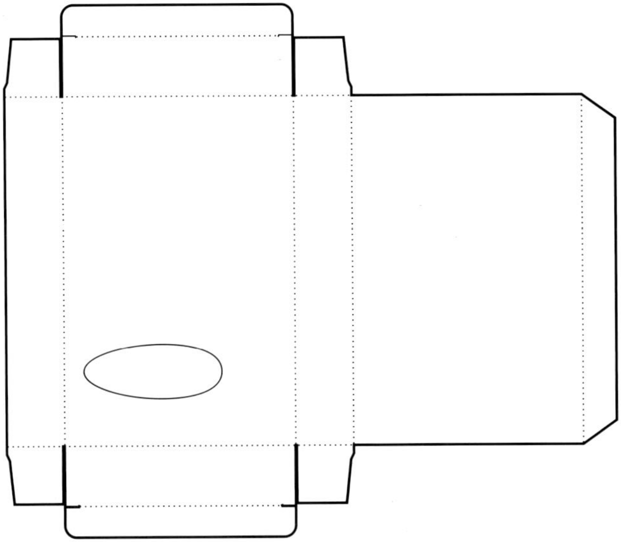

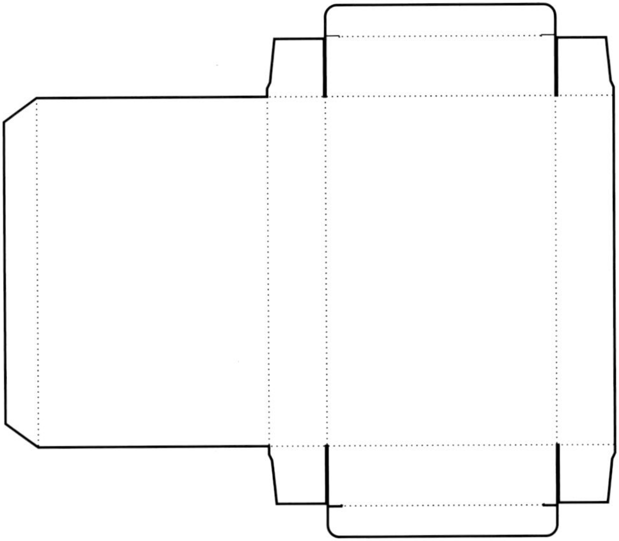

BRAUN
New
Twin Control
4605
Twin Foil
Rechargeable
Washable
4000 series

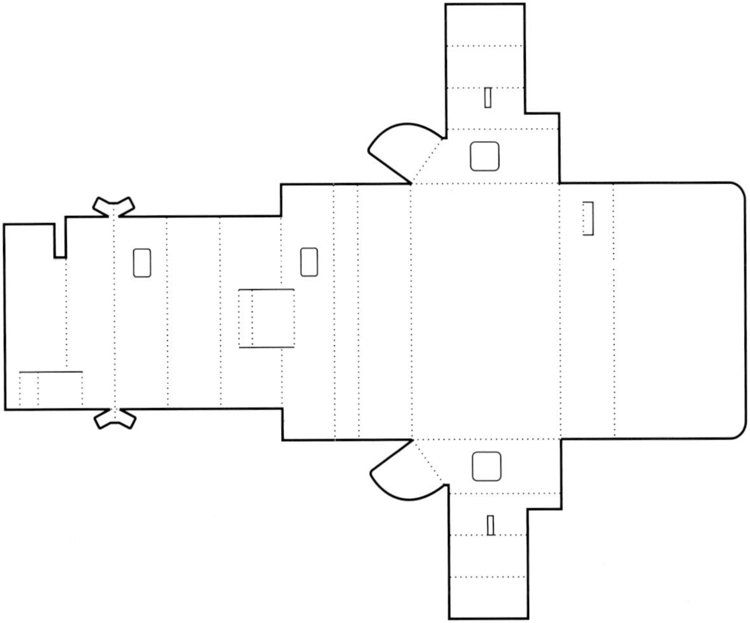

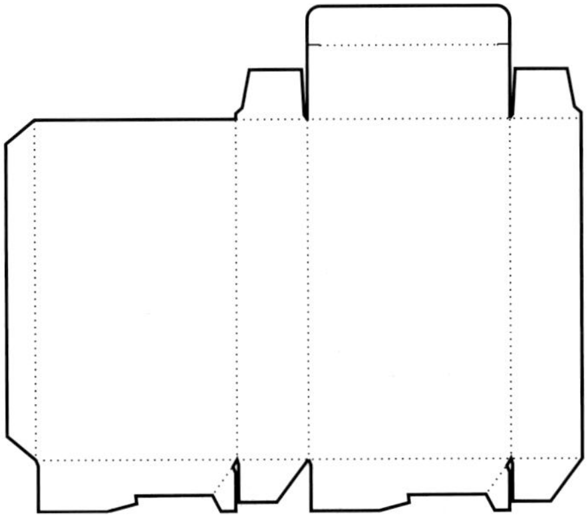

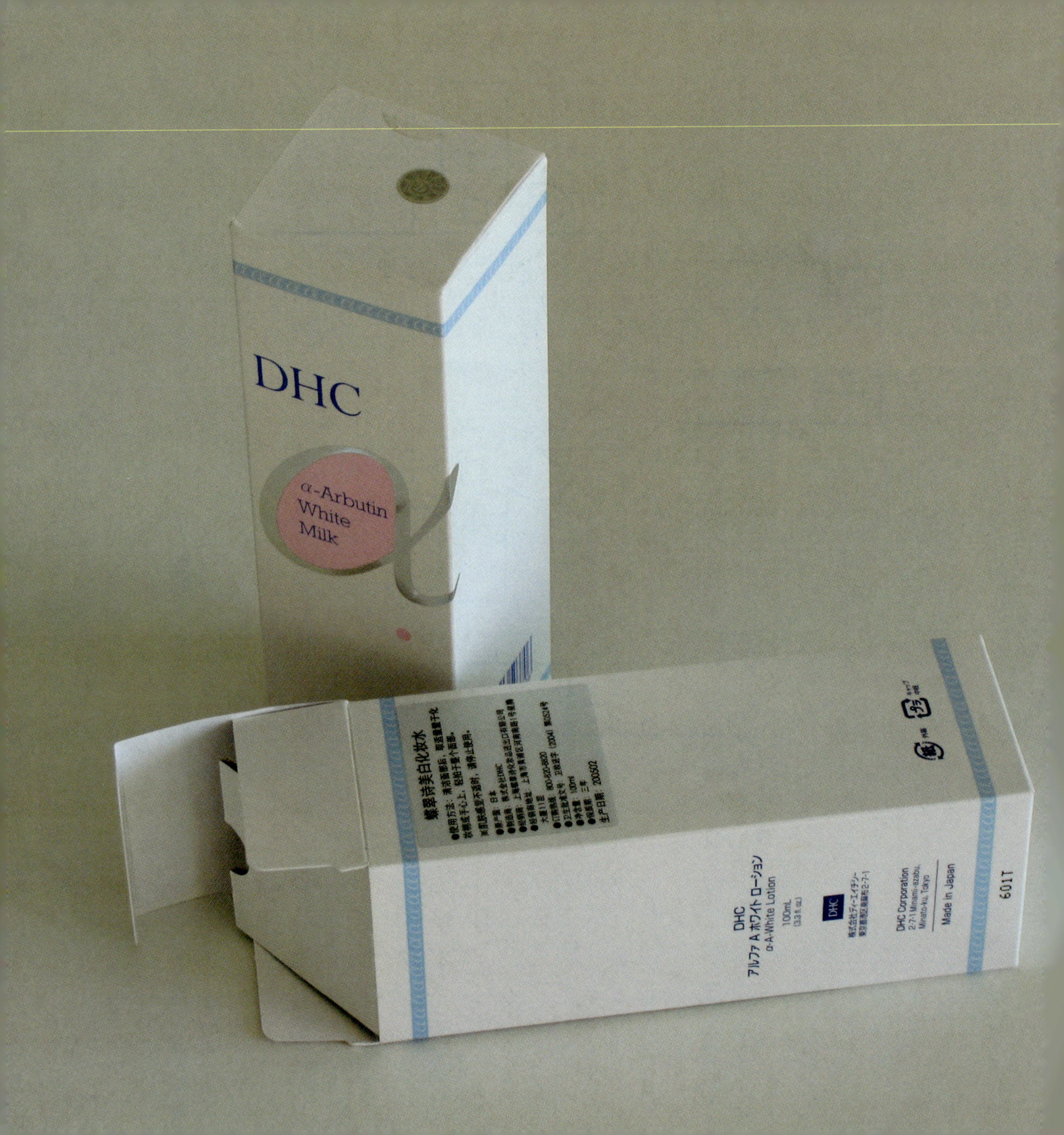
DHC
α-Arbutin
White
Milk

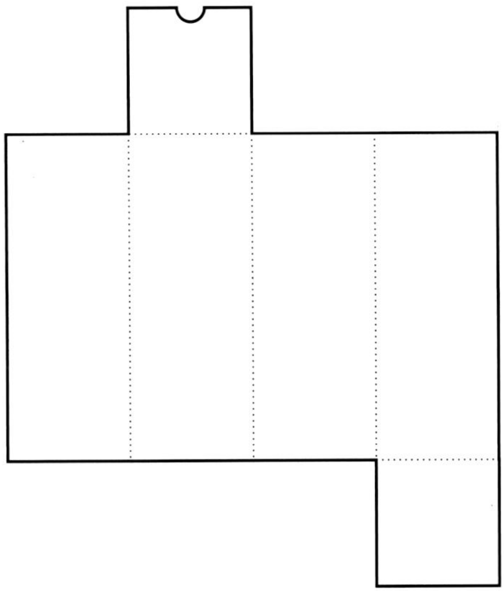

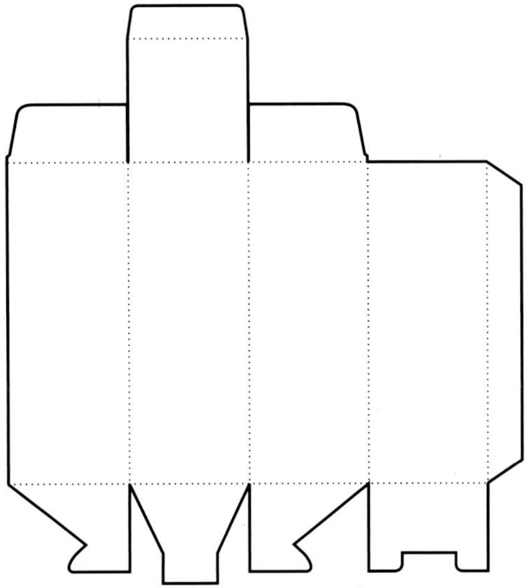

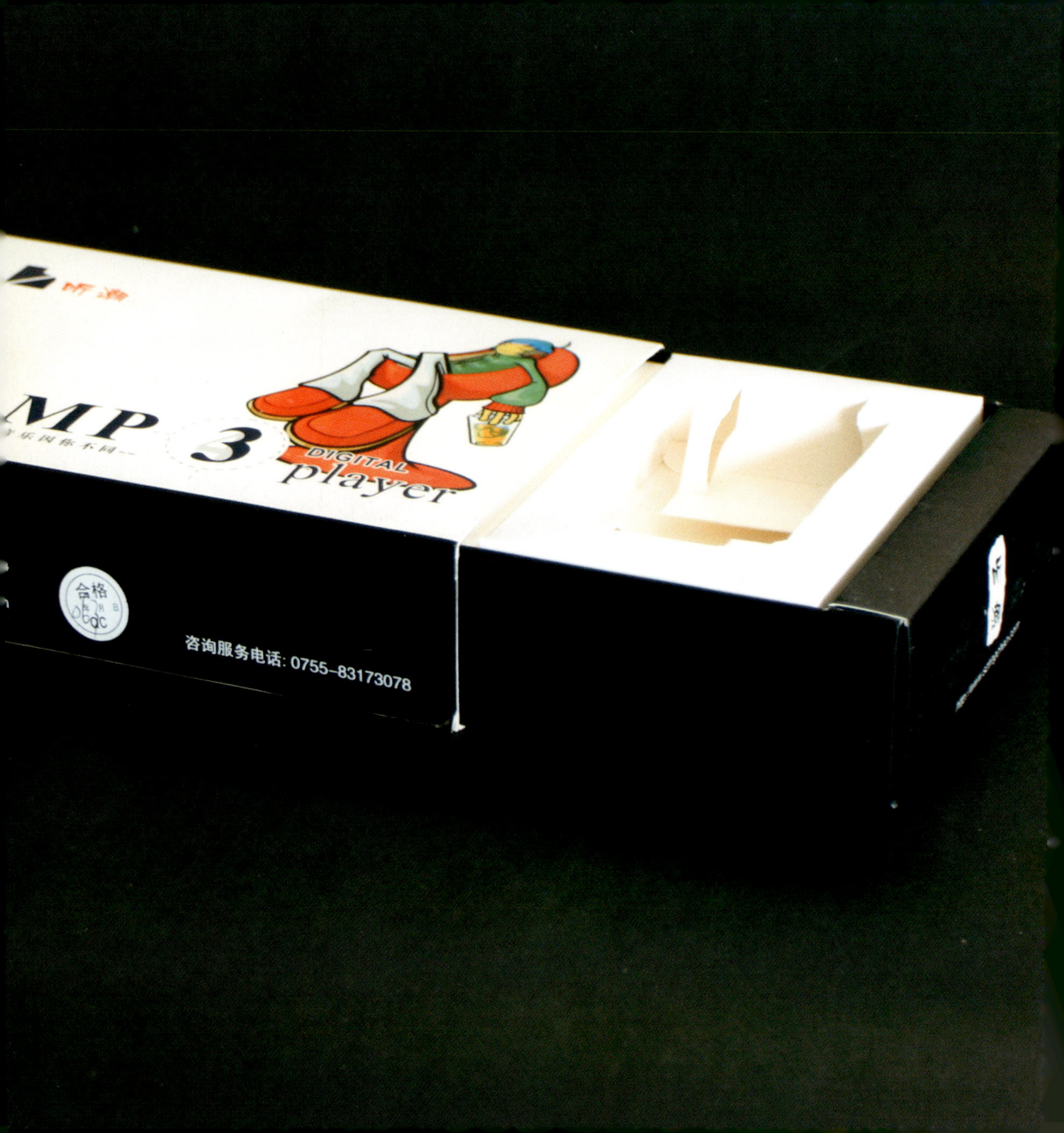
MP 3
DIGITAL
player
咨询服务电话: 0755-83173078
合格

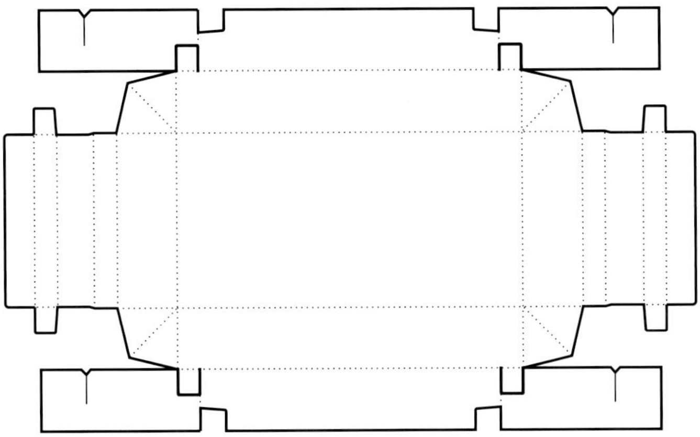

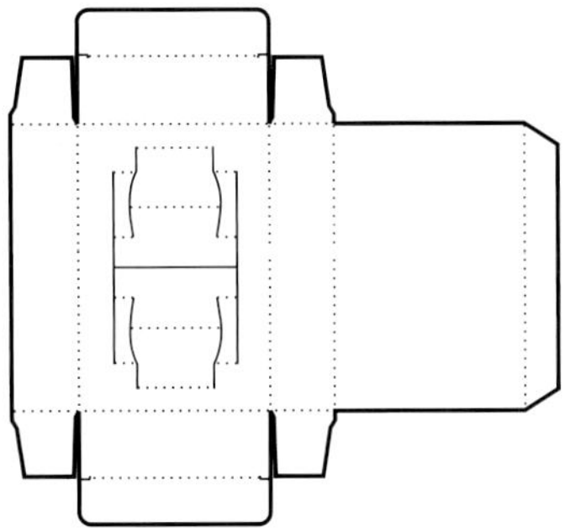

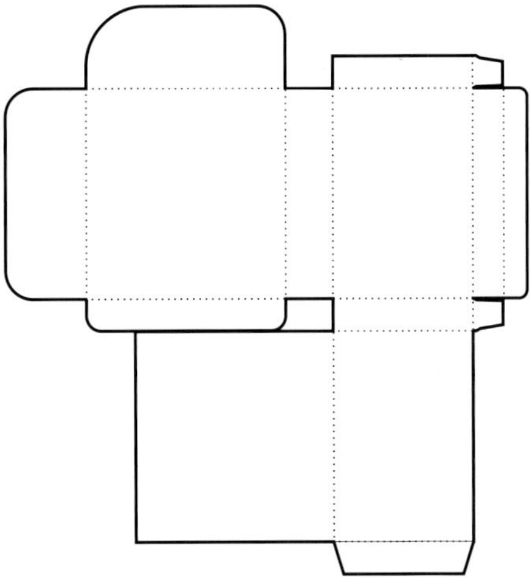

ALpha
阿尔法
专为在线游戏优化和家庭共享上网设计

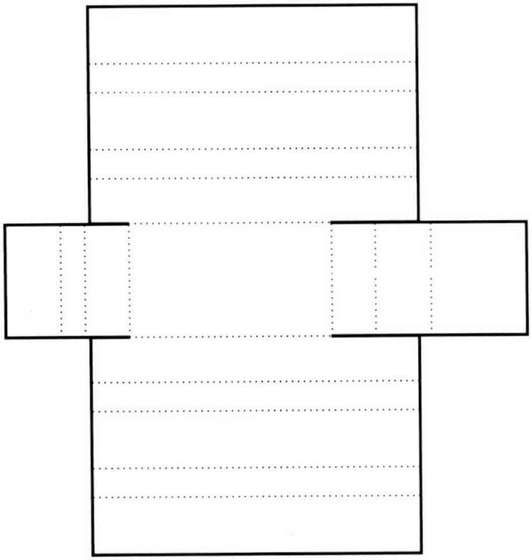

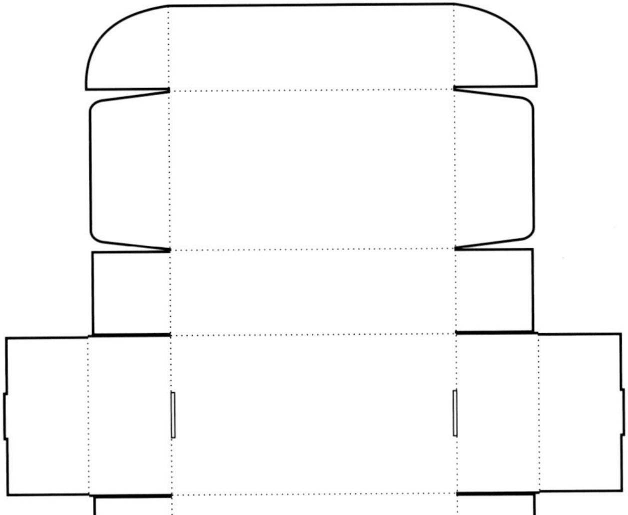

魔鬼眼
GHOST EYES
独具微光夜视功能
数码摄像头

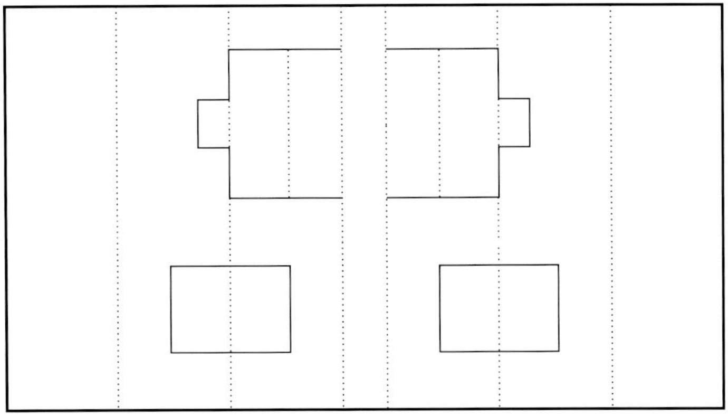

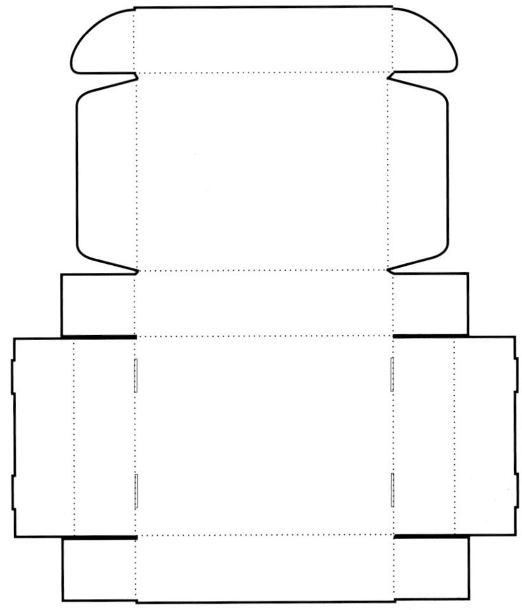

Light
more
Pure
more
iMAX
www.imax.com.cn

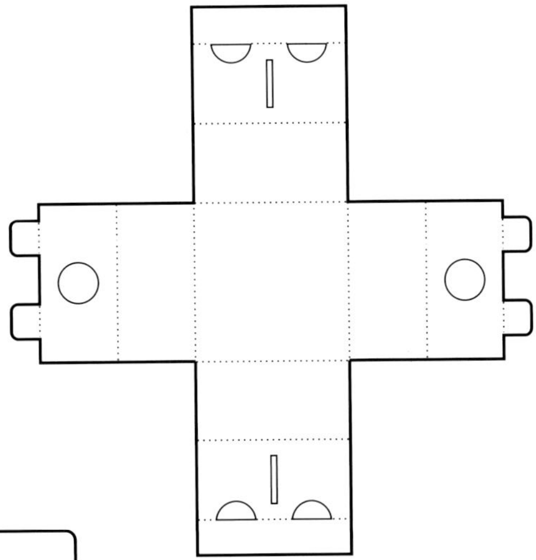

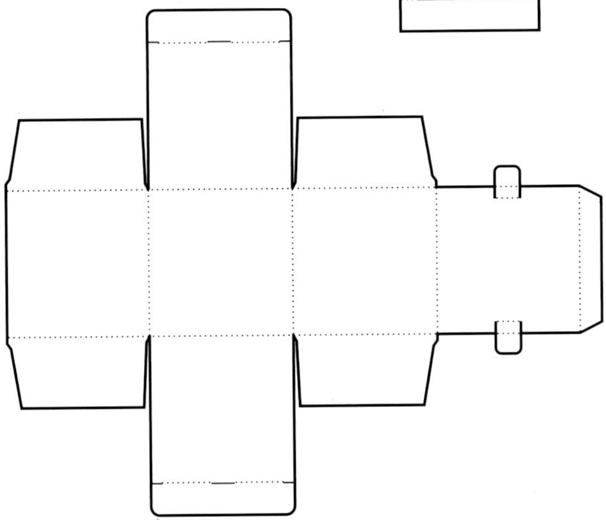

NOKIA

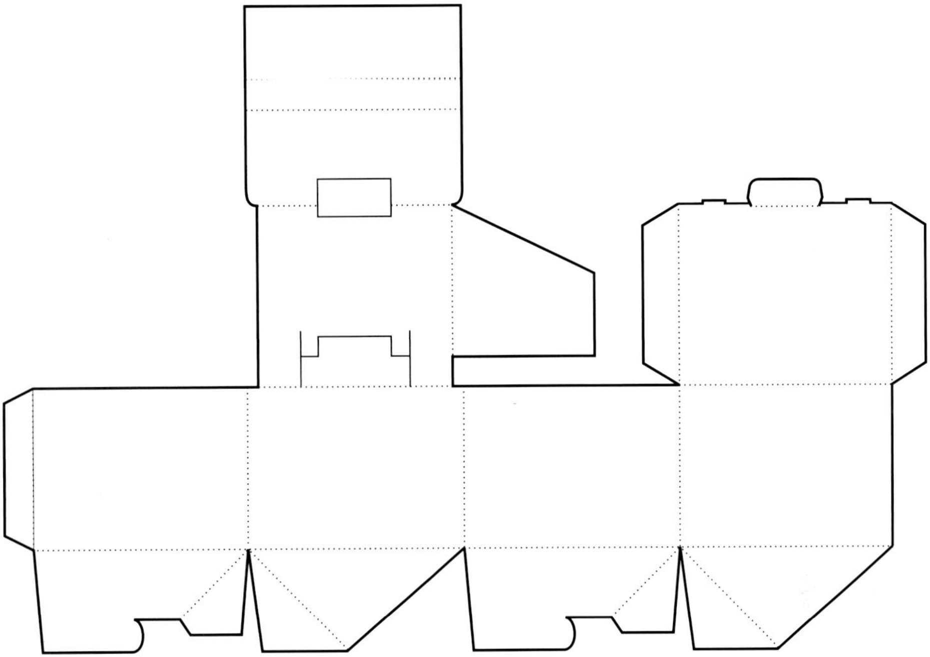

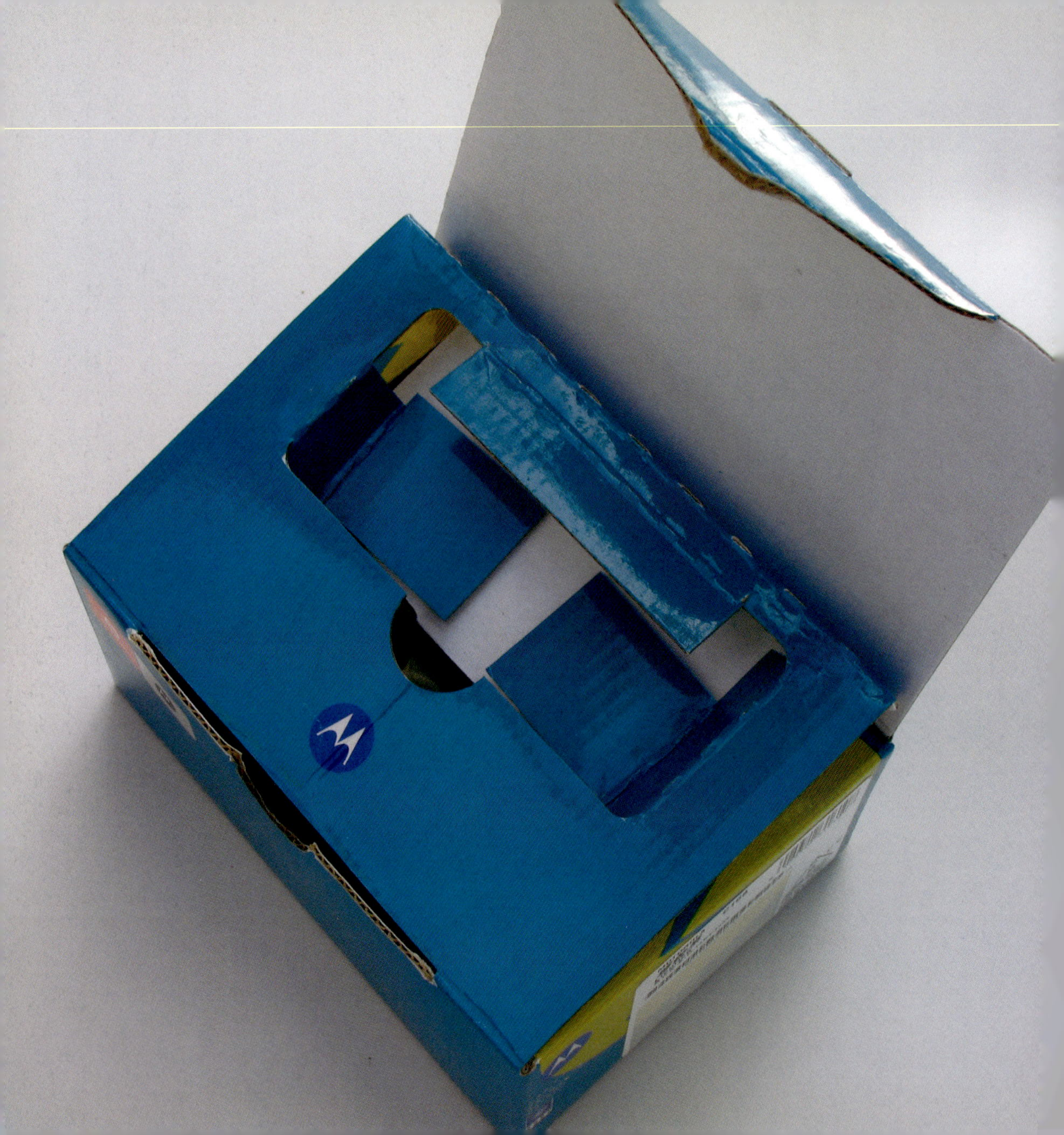

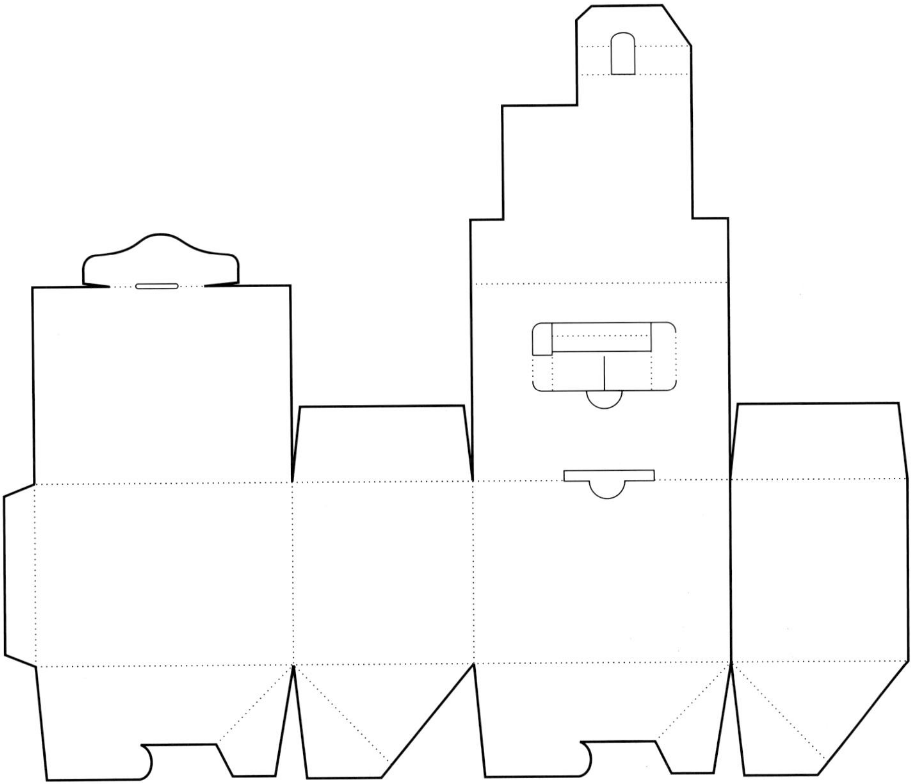

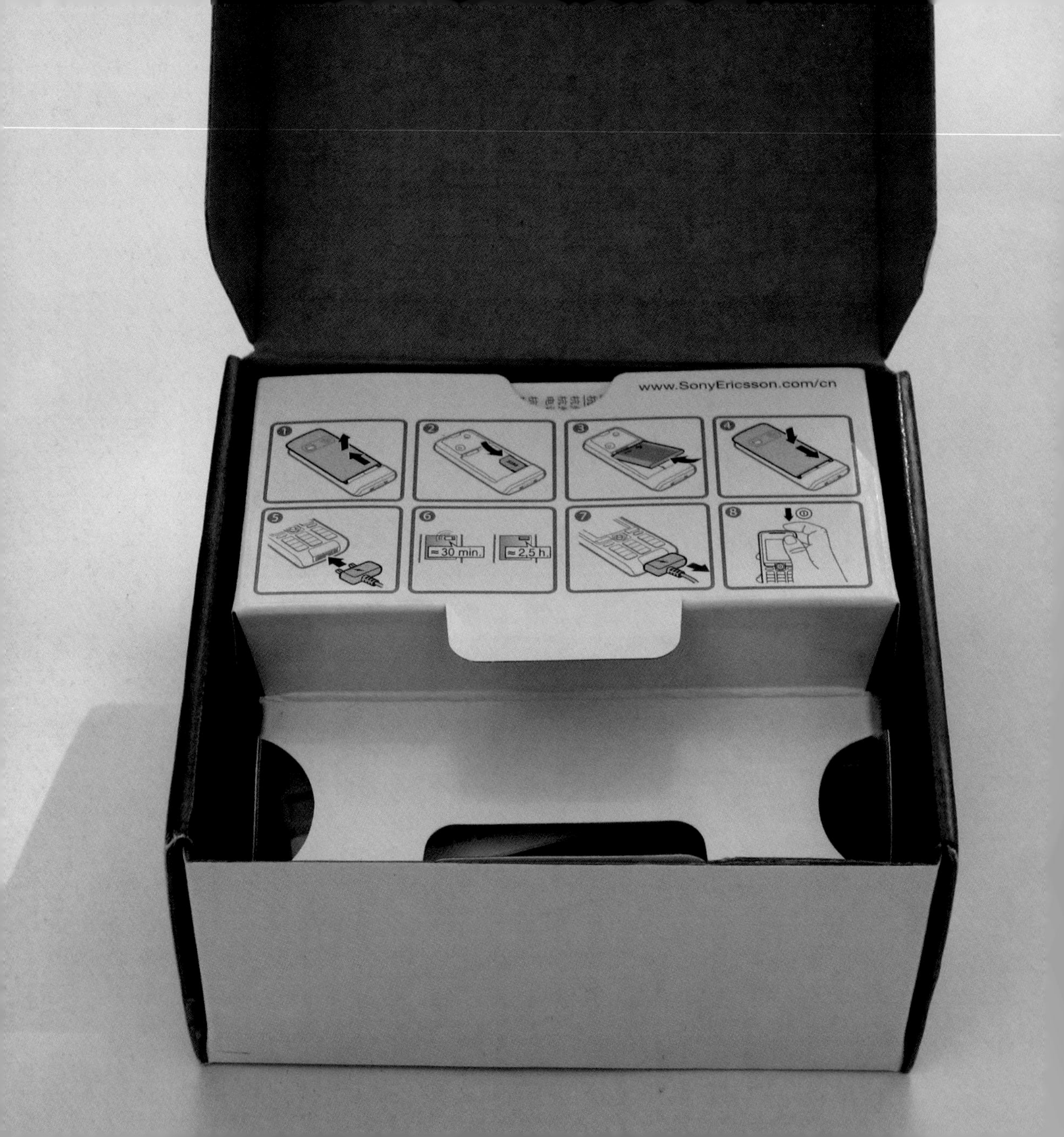
www.SonyEricsson.com/cn
≈30 min.
≈2,5 h.

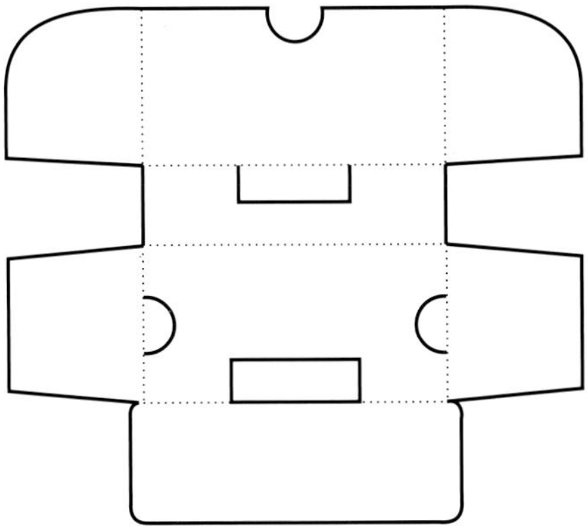

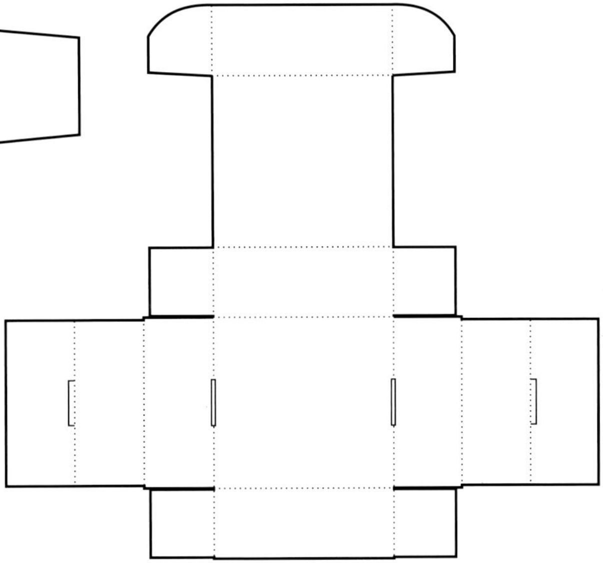

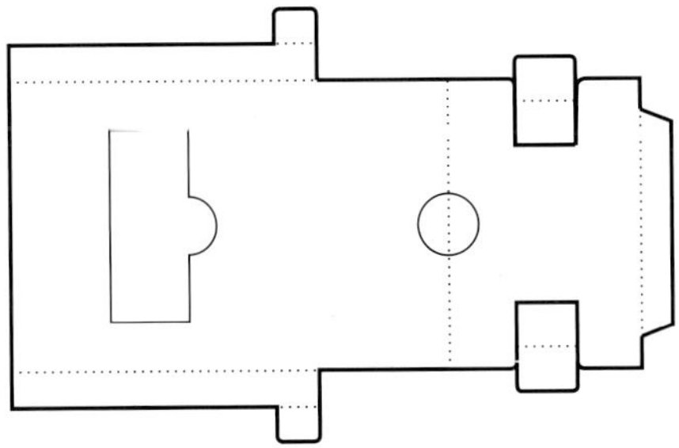
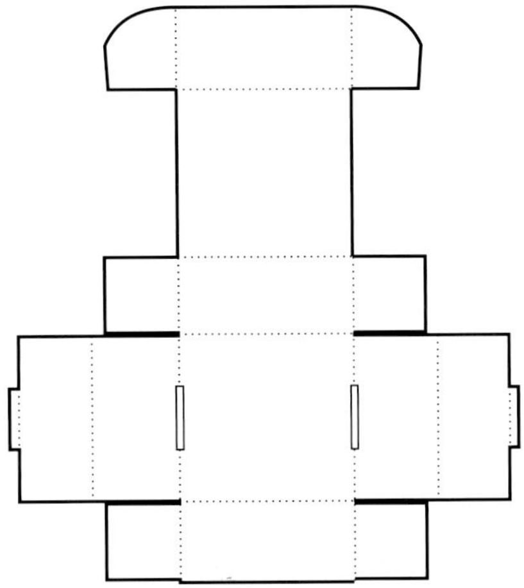
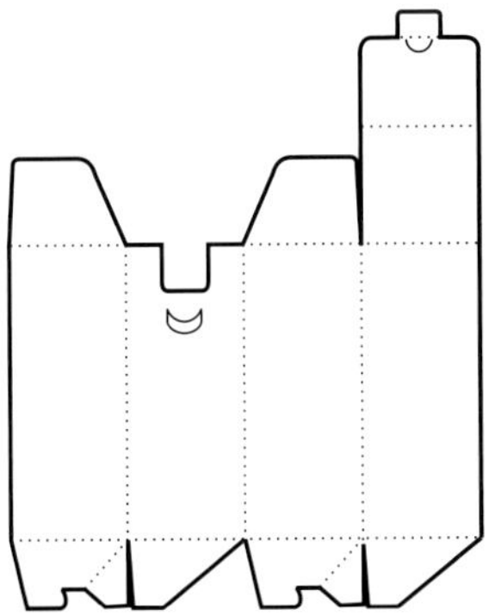
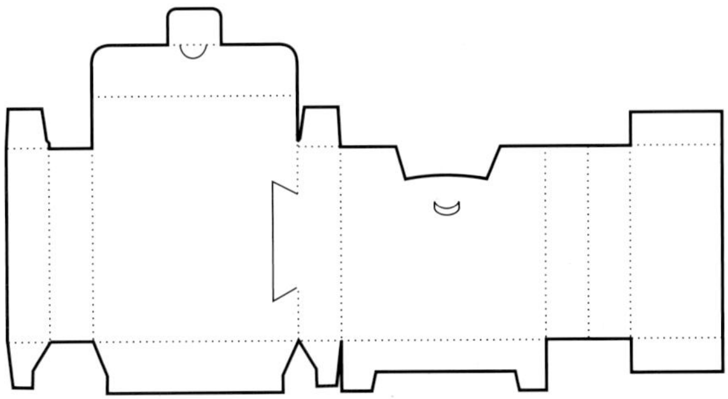

SONY

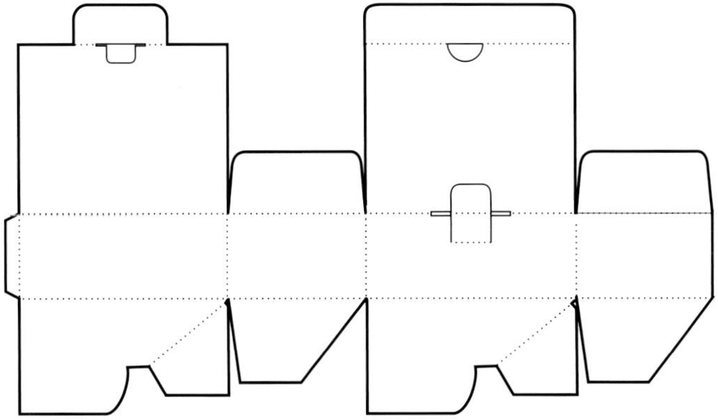

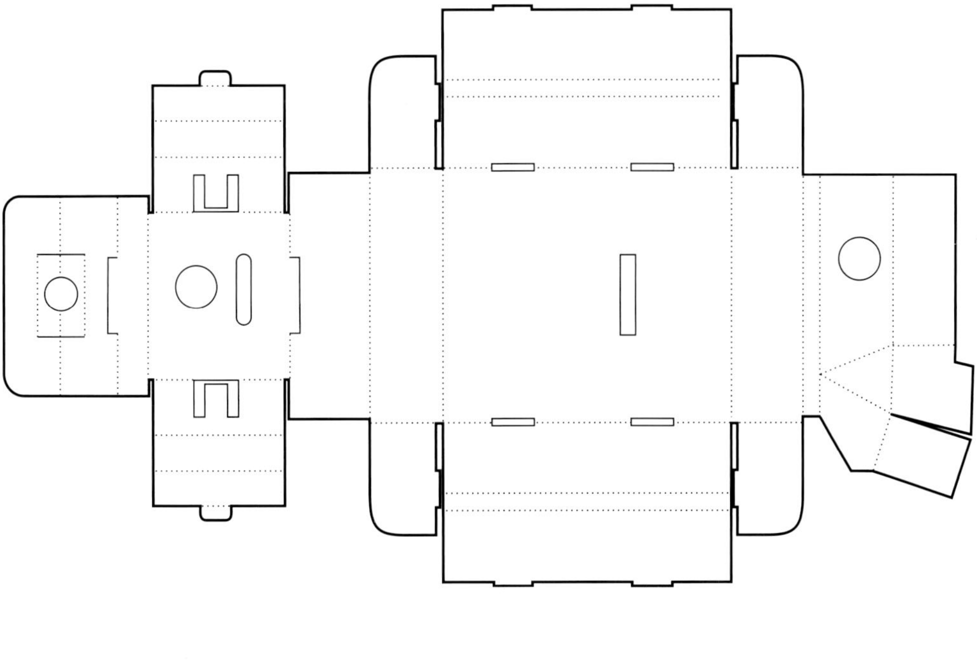

TWIN® I.V.I
TWIN® I.V.I
Perfect ergonomics, perfect control
TWIN® I.V.I

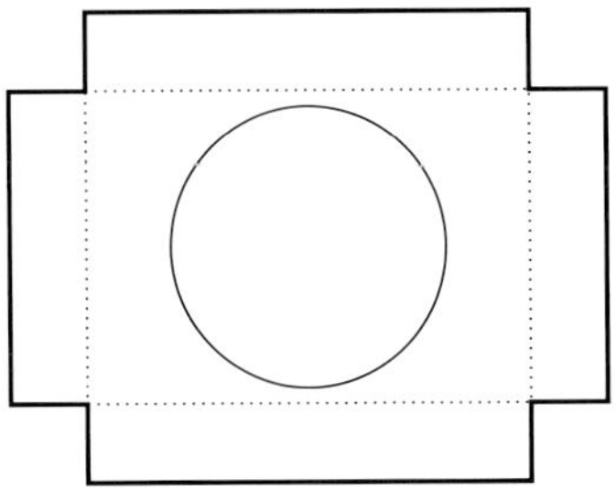

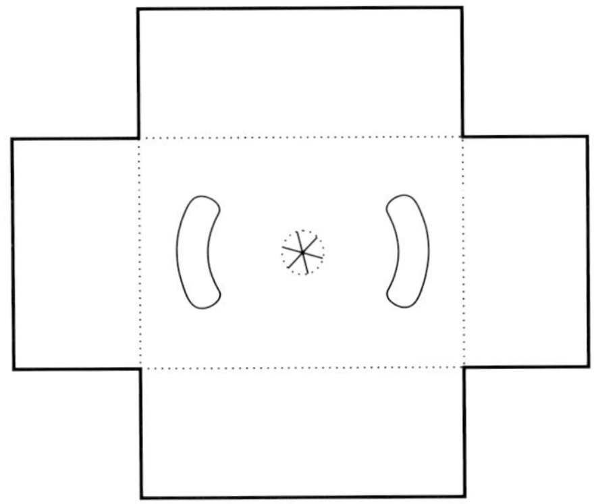

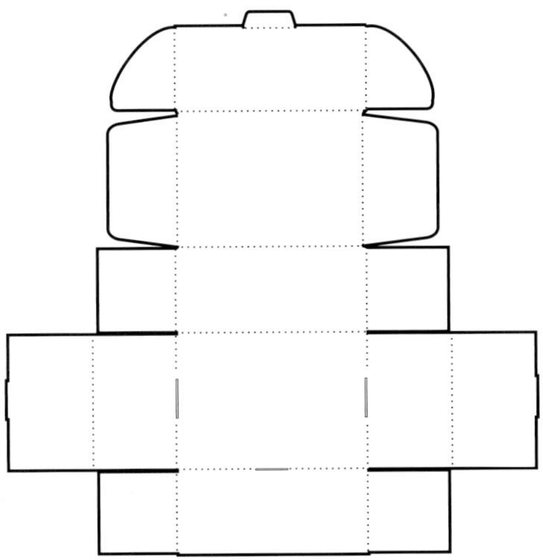

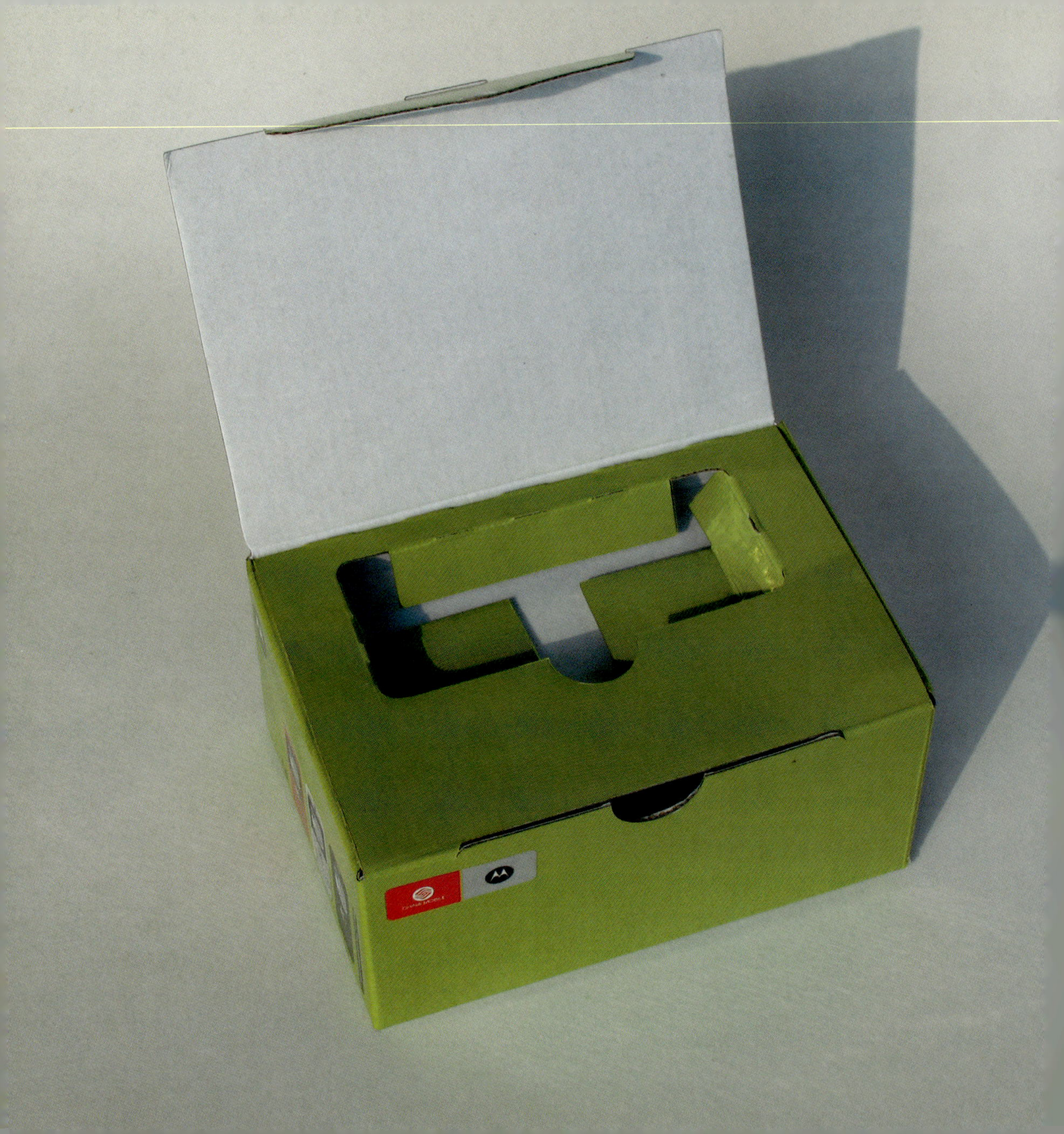

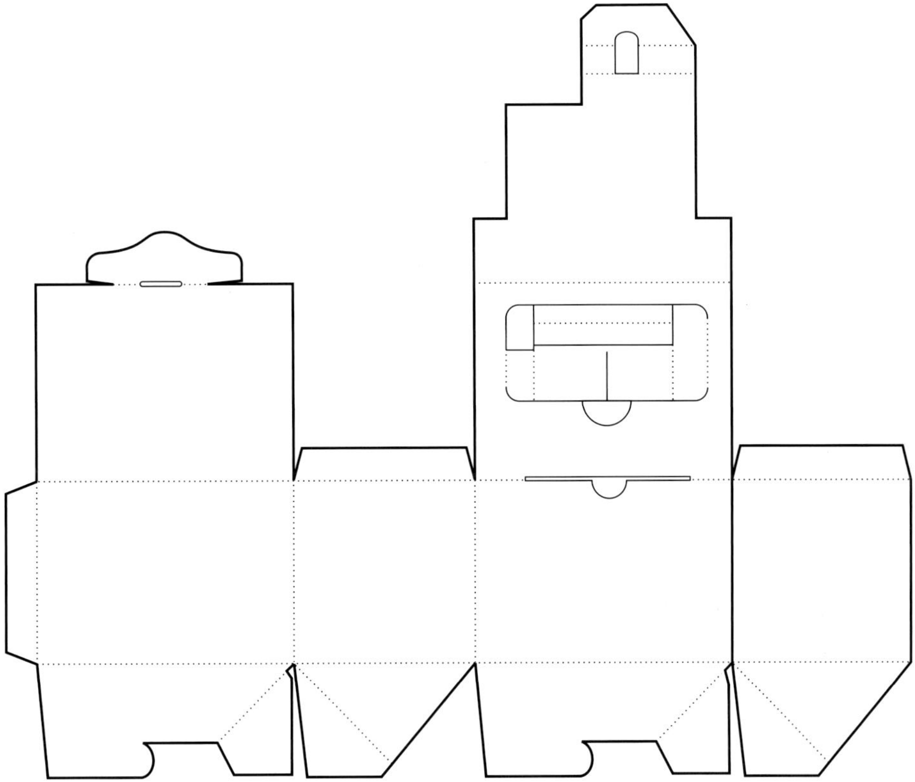

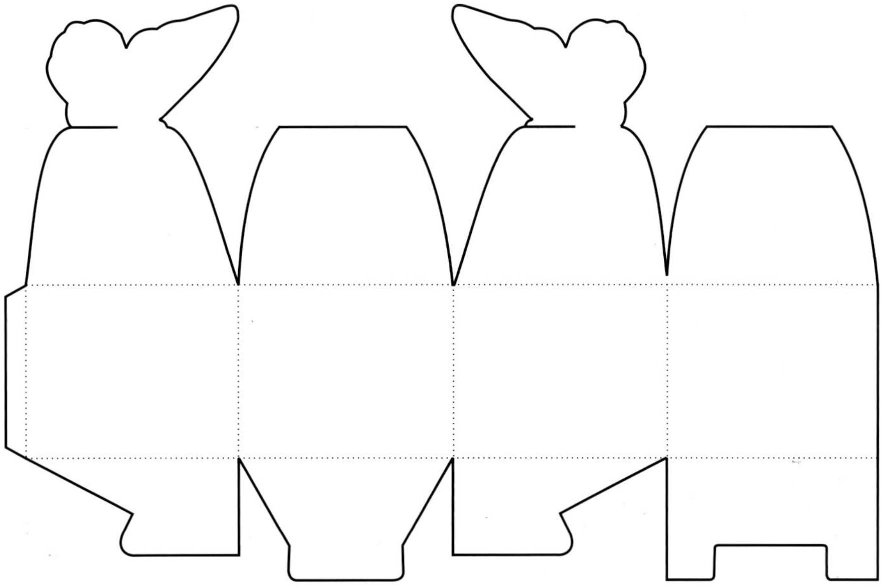

阿尔卑斯
草莓牛奶硬糖

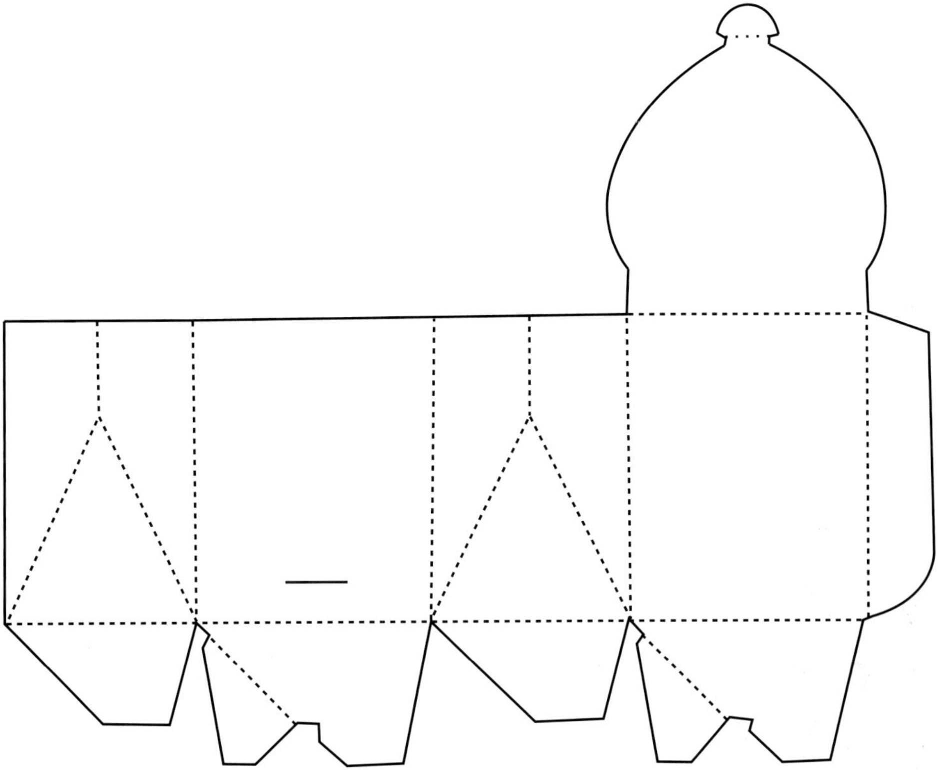

花嫁喜鋪

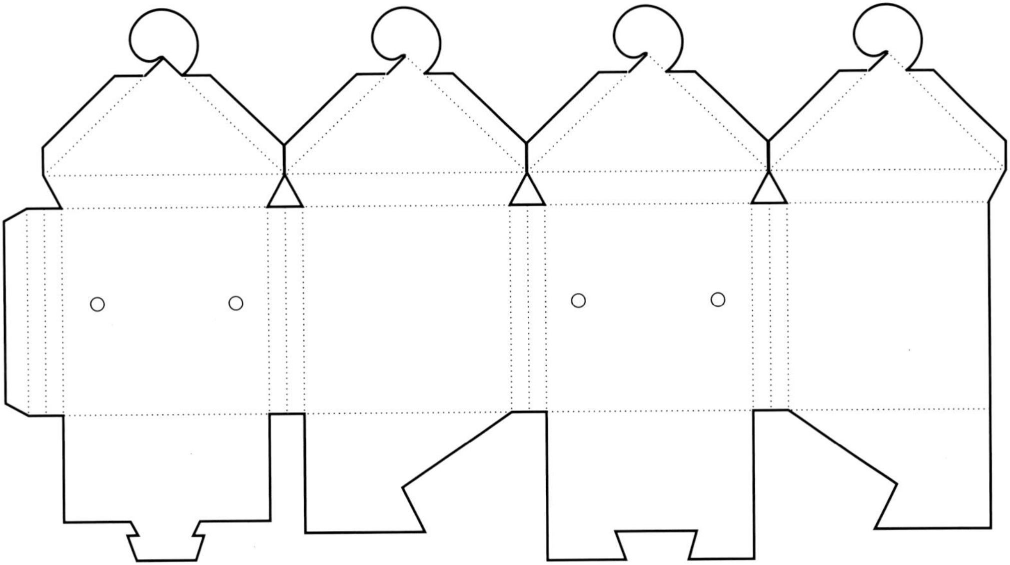

小口微
水果风
FRUITS DIVERS EN VOGUE
瑞士糖
RUI SHI TANG
100%草莓口味
STRAWBERRY FLAVOR

FANGLAIYA XIAOQIMAO
宝宝肌肤
细心呵护
最新一代
FANGLAIYA XIAOQIMAO
宝宝肌肤
细心呵护
儿童柠檬润肤霜
Children's Skin Cream with Lemon

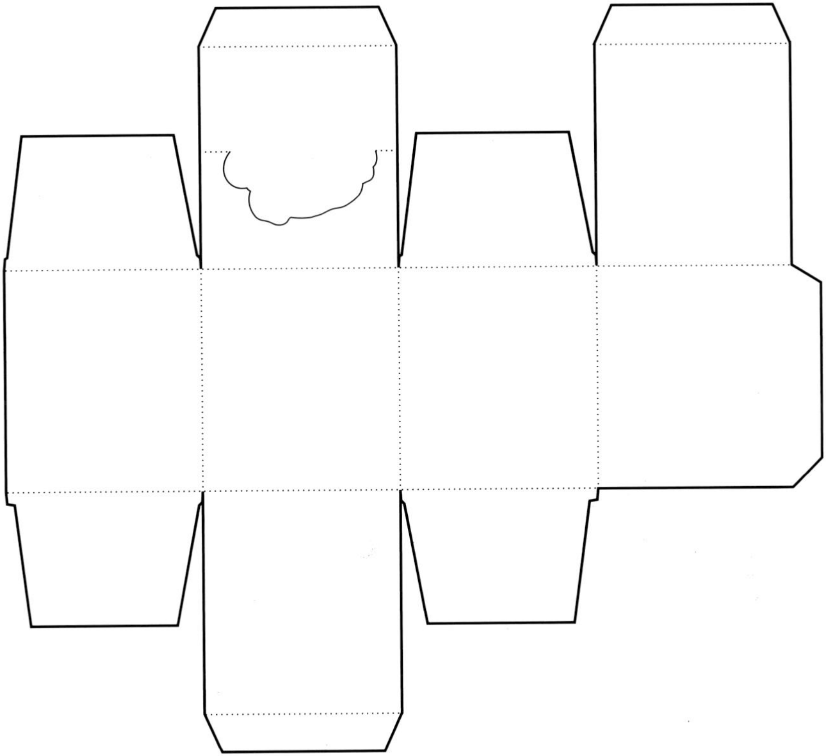

FLOWER
&
HEYB
玫瑰
丰盈洗发乳
Rose
Nourishing Shampoo
净含量：250g

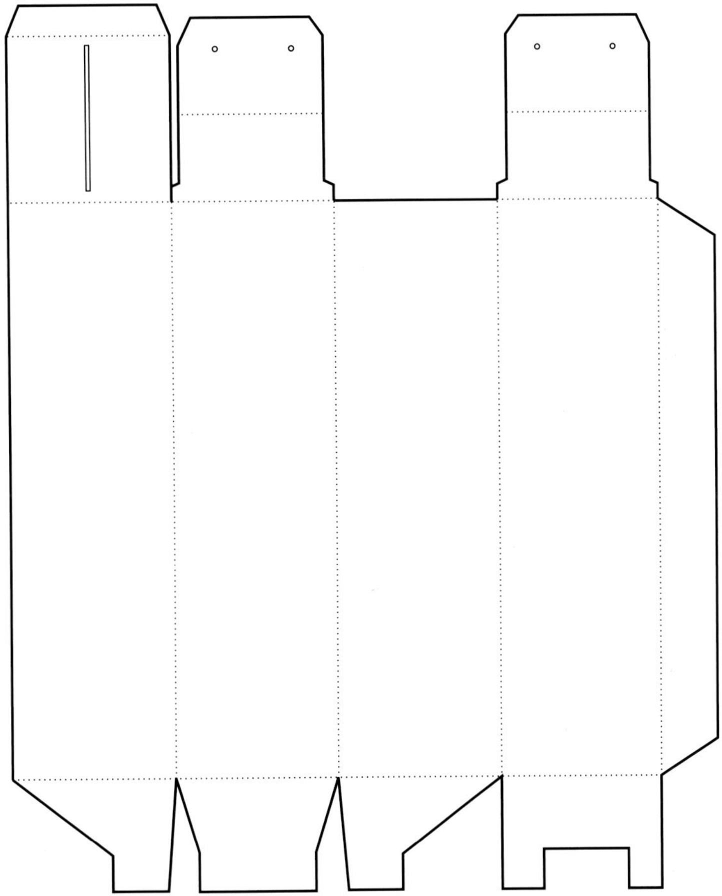

同心结
奶油巧克力喜糖
NAIYOU CHOCOLATE
TONGXIN JIE XITANG CHOCOLATE
同心结
奶油巧克力喜糖
NAIYOU CHOCOLATE
TONGXIN JIE XITANG CHOCOLATE

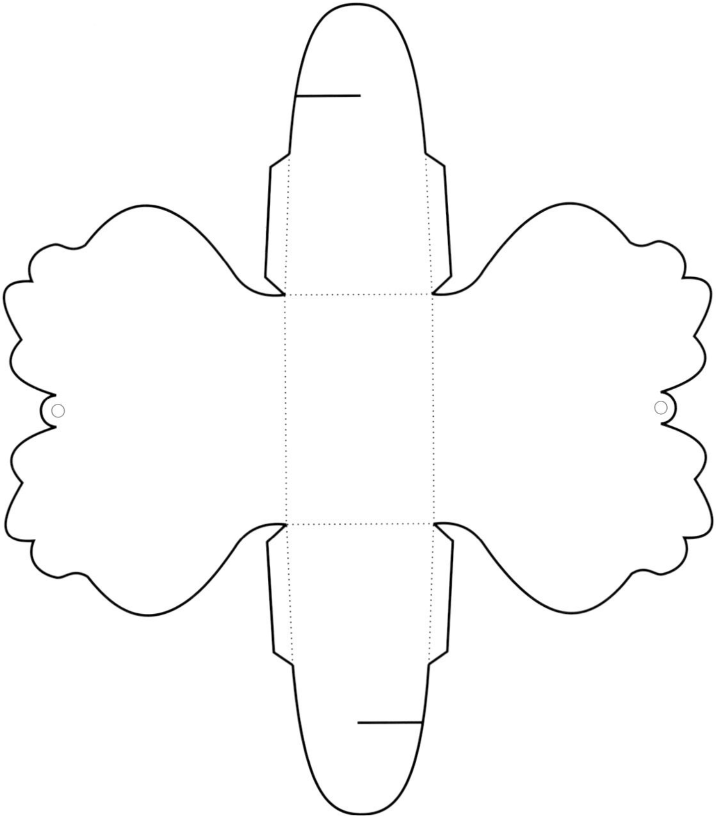

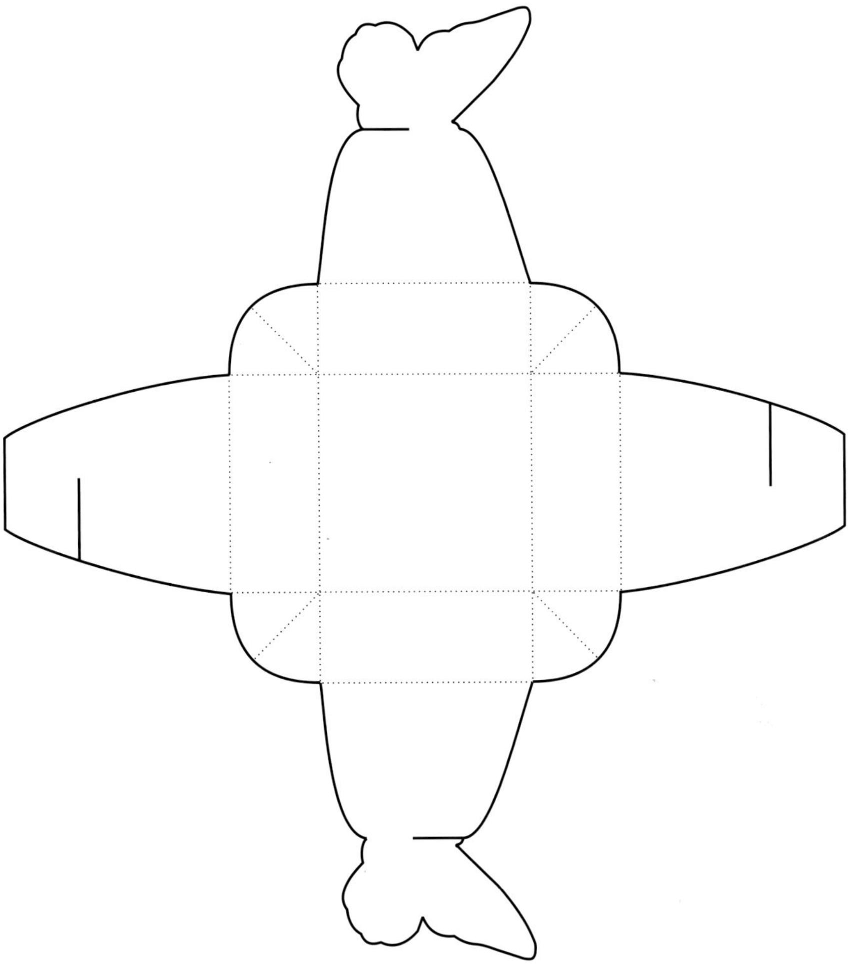

Dove
德芙
越洋奇珍
一夹心黑巧克力
Silky Dark Truffle
净含量:60g

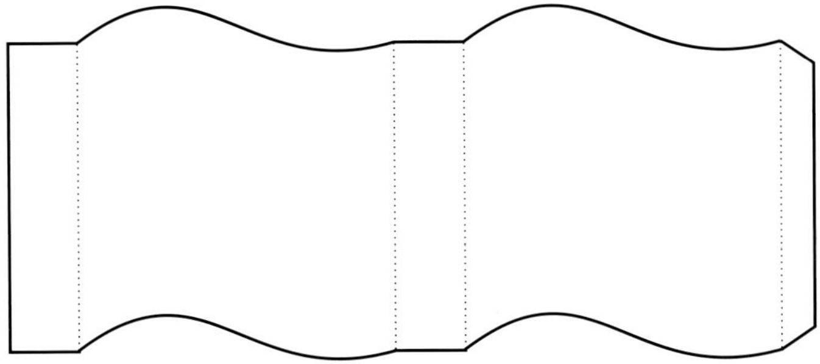

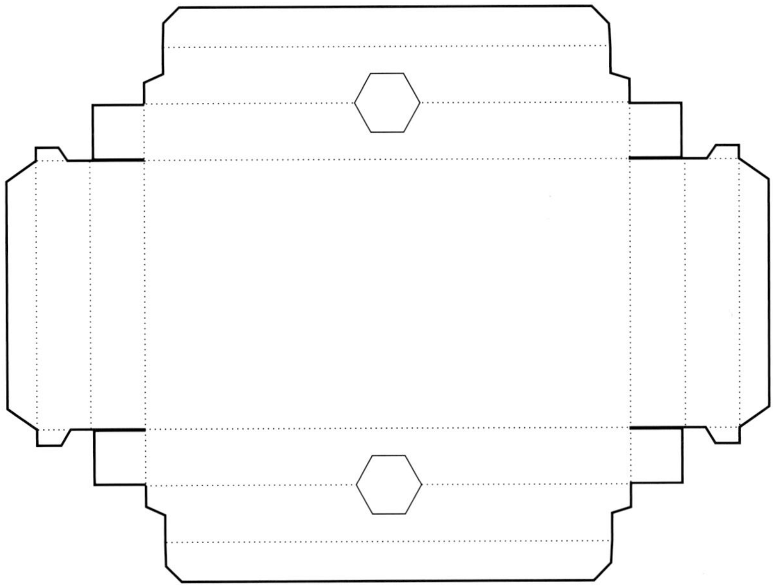

LOVE

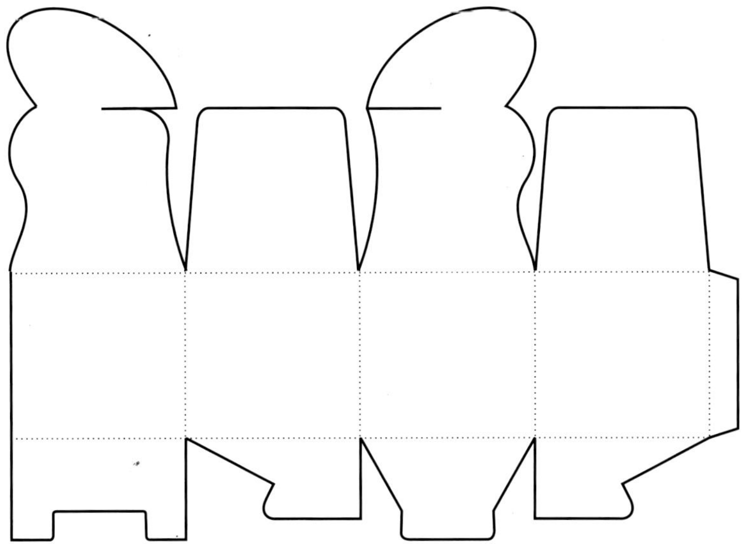

诗蒂®
Shinetien
With Love
诗蒂®
Shinetien
With Love

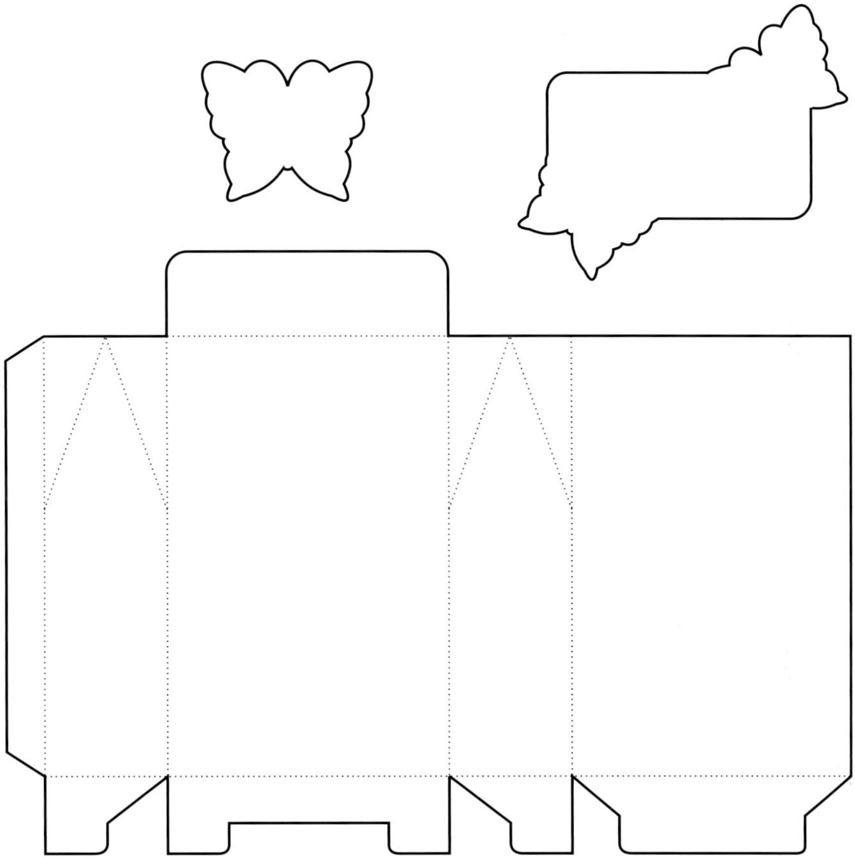

伴侣巧克力

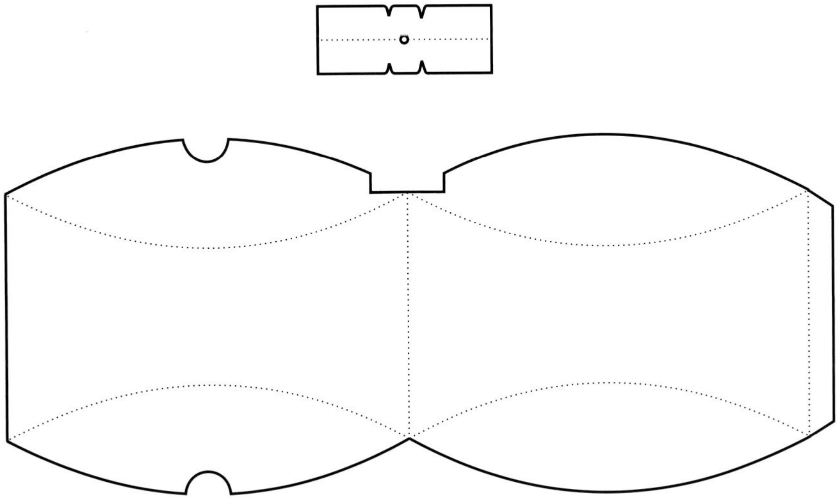

我的主題喜糖
iSugar

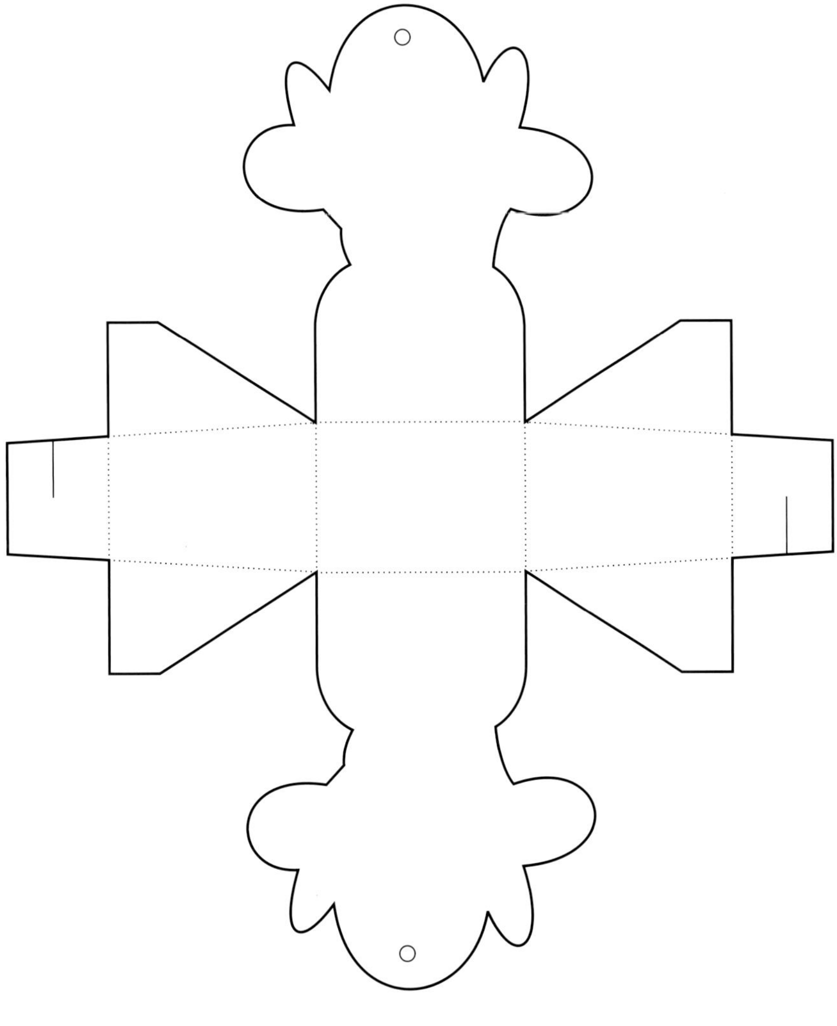

嫁給我
嫁給我

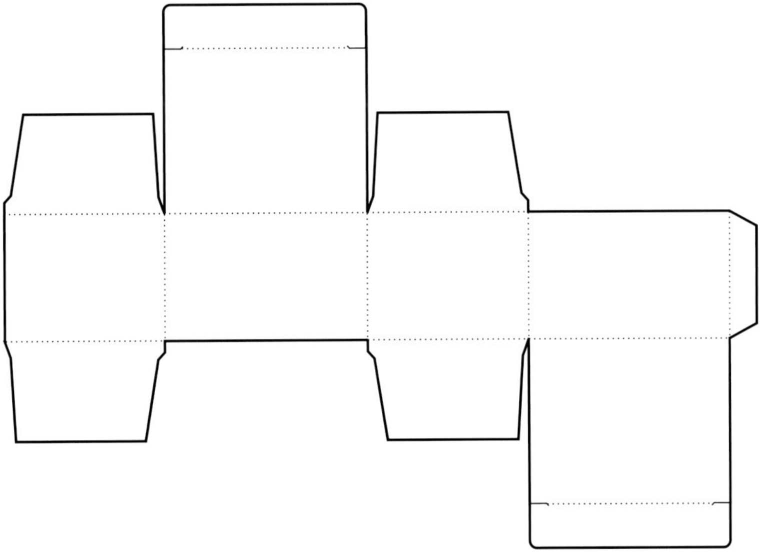

百年
好合
缔结
良缘
百年
好合

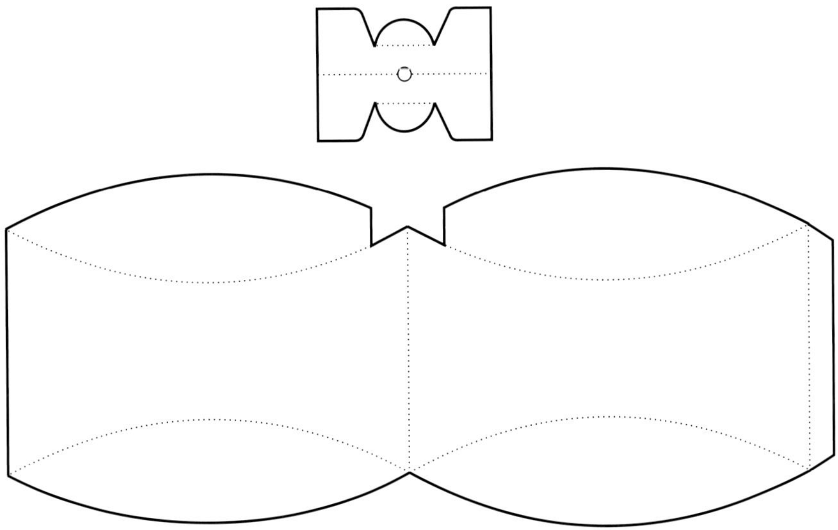

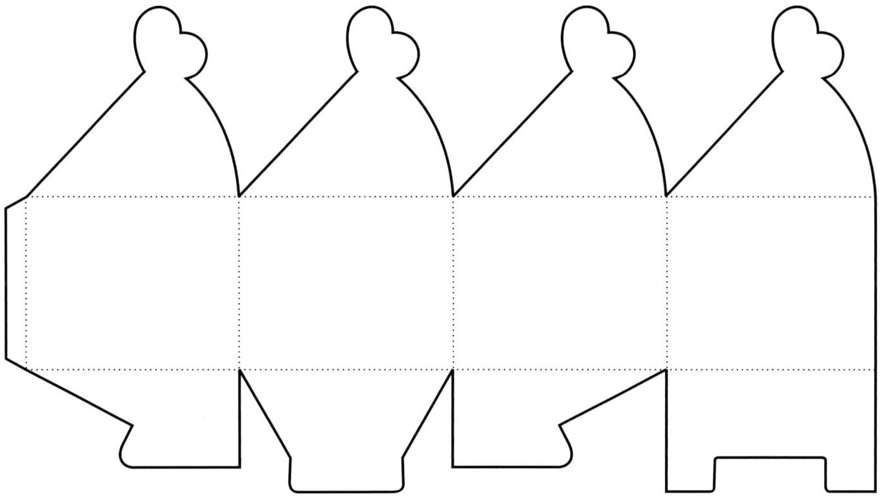

XiZhiLian
喜之恋
喜糖

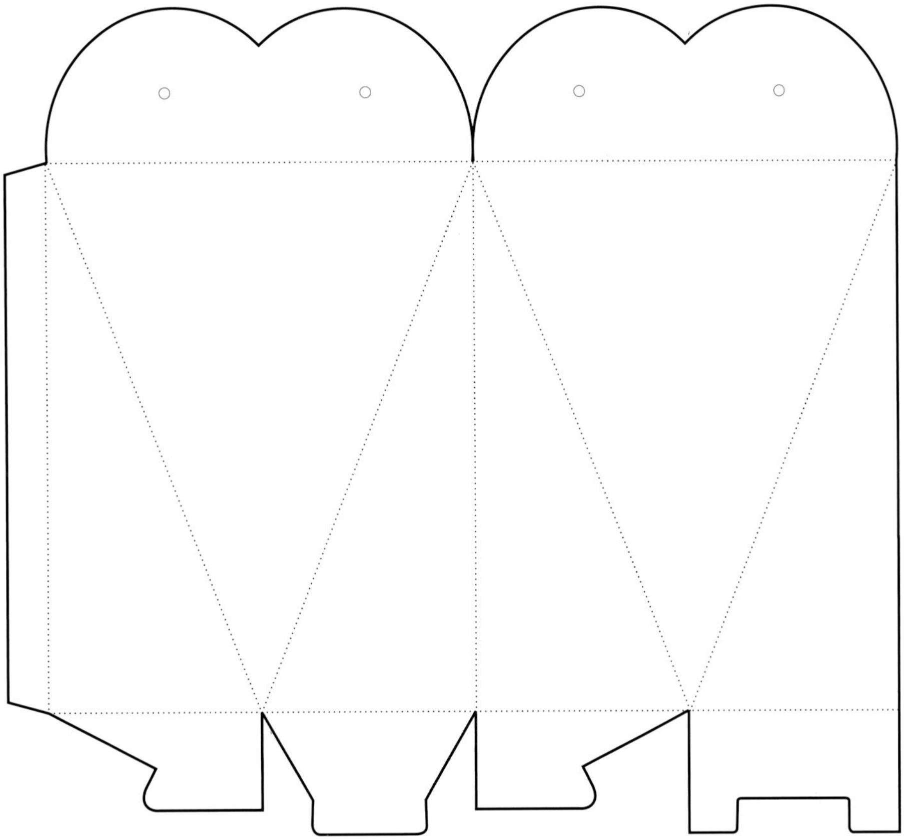

中秋月餅
MOON CAKE

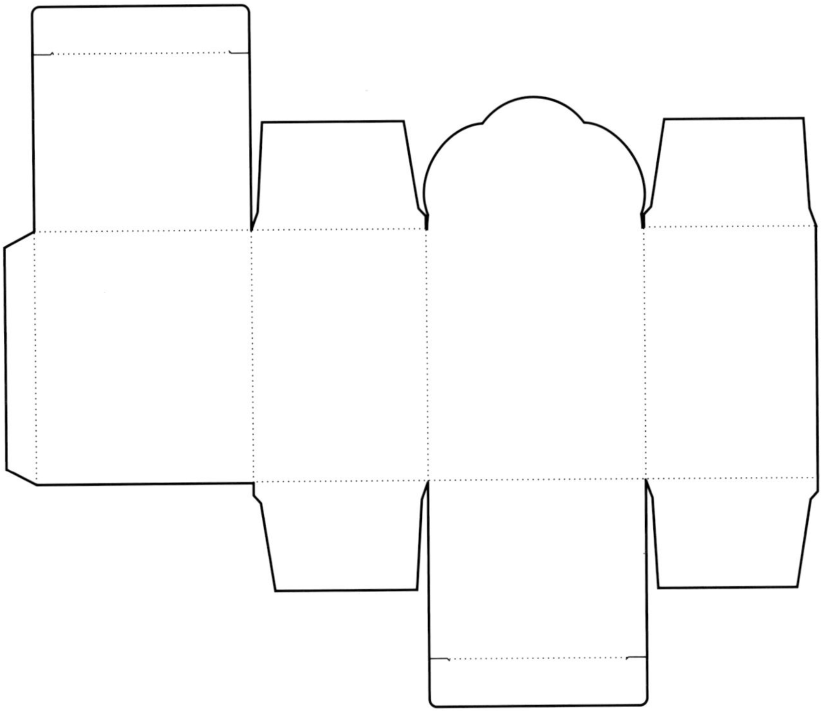

相爱一生
情诗
Qing Shi
Qing Shi
相爱一生
XIANG AI
YI SHENG

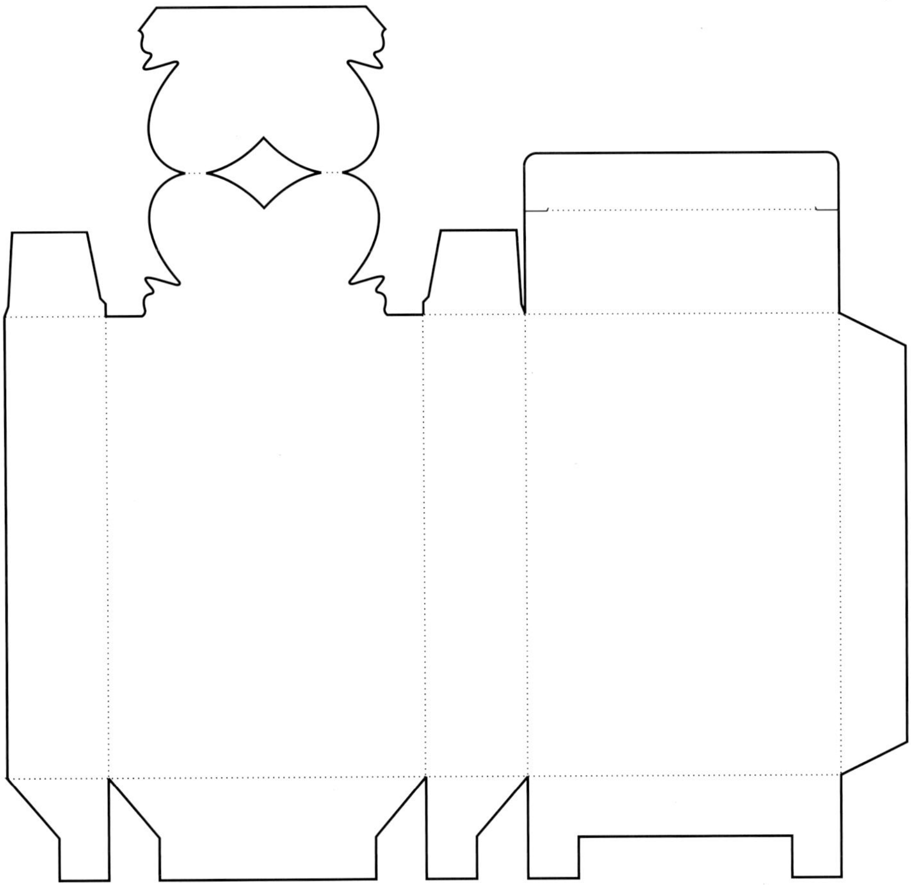

Welcome
to our wedding!
诗蒂
Shinetien

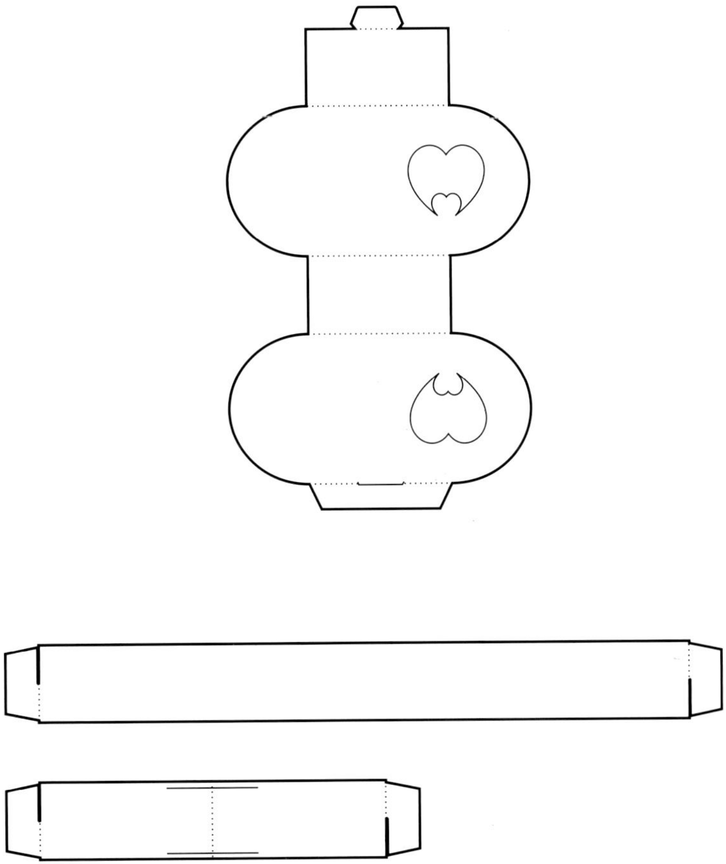

冬蓉
嬌滴滴
老婆餅
WING WAH
WING WAH
Since 1950
Cutie Wife Cake
sweet flaky pastry
Chinese Delicacies With
A Timeless Flavour

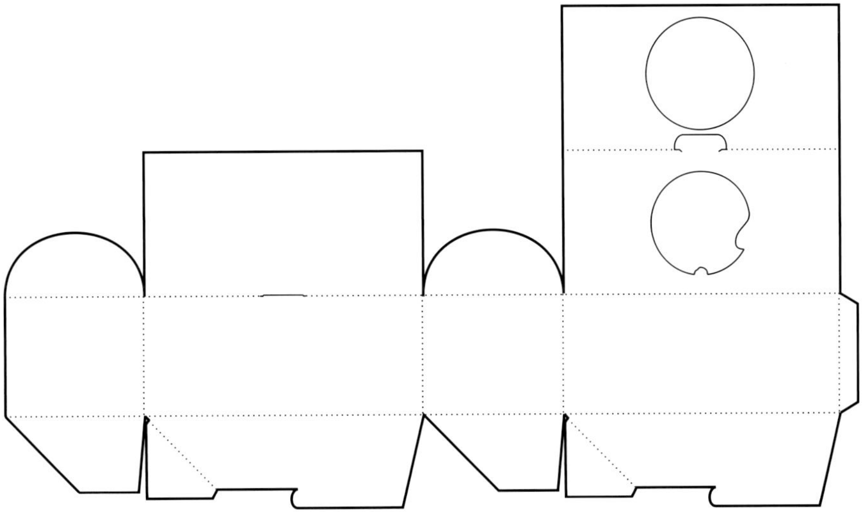

MOON CAKE
— 月饼 —
MOON CAKE

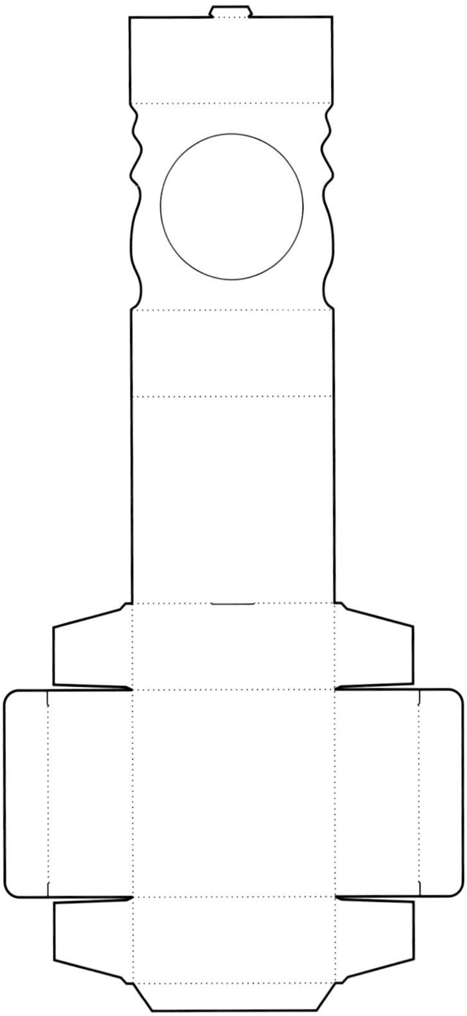

诗蒂
Shinetien

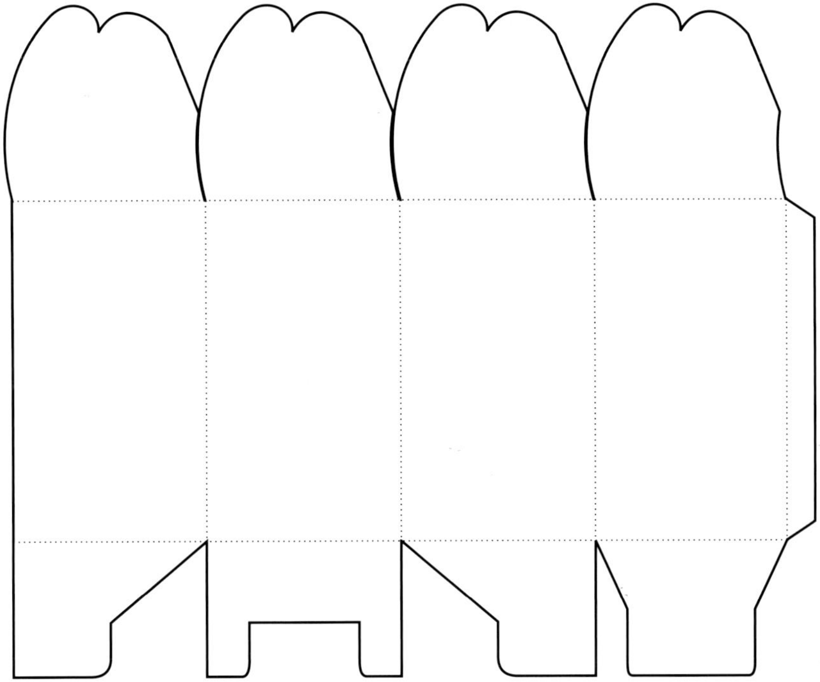

囍
阿尔卑斯®
正宗
特制纯真 高级牛奶糖

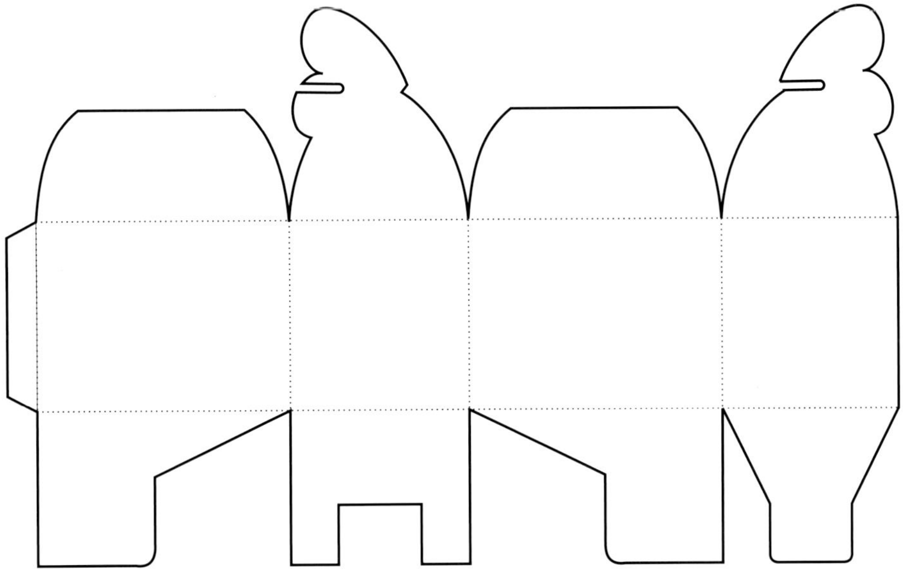

天赐良缘
永结同心

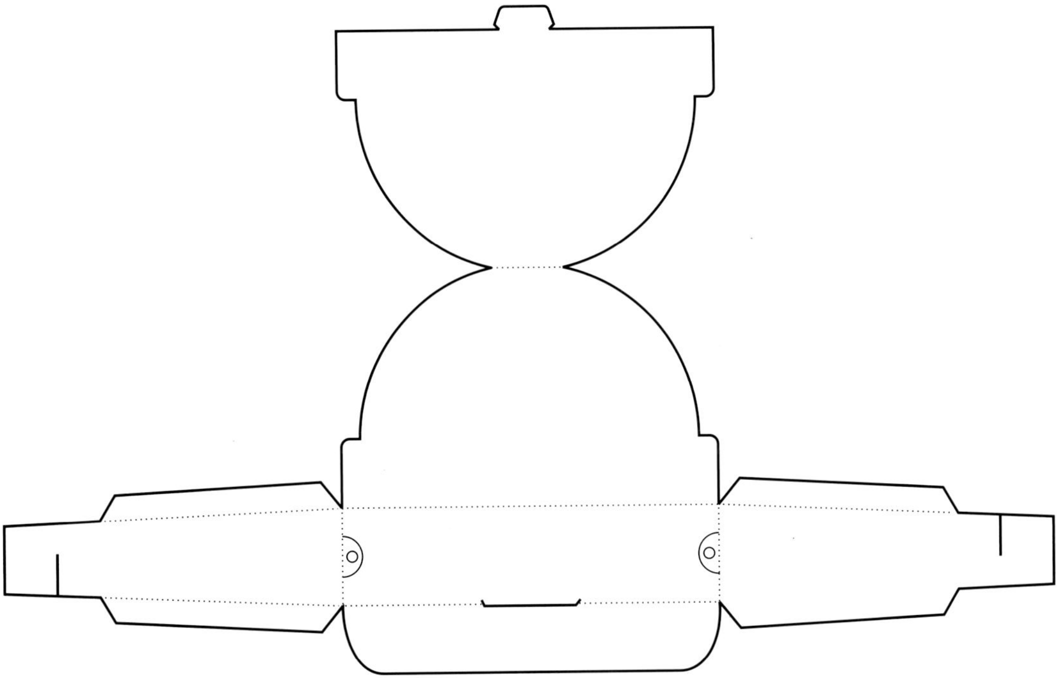

浓情
蜜意
锦绣良缘
Nongqingmiyi
Nongqingmiyi

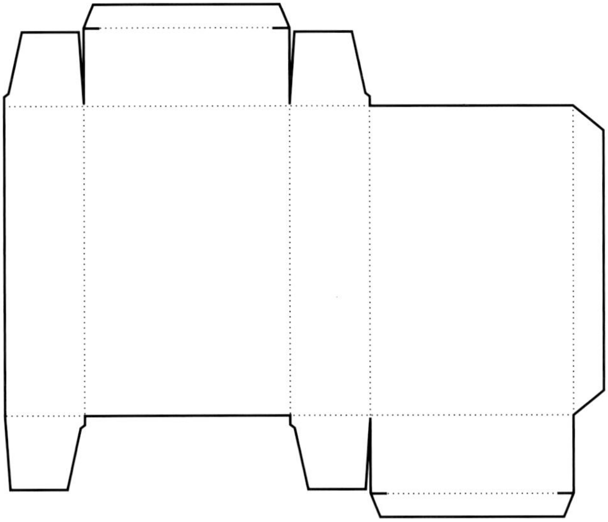

百年好合
永结同心
Golden Bird

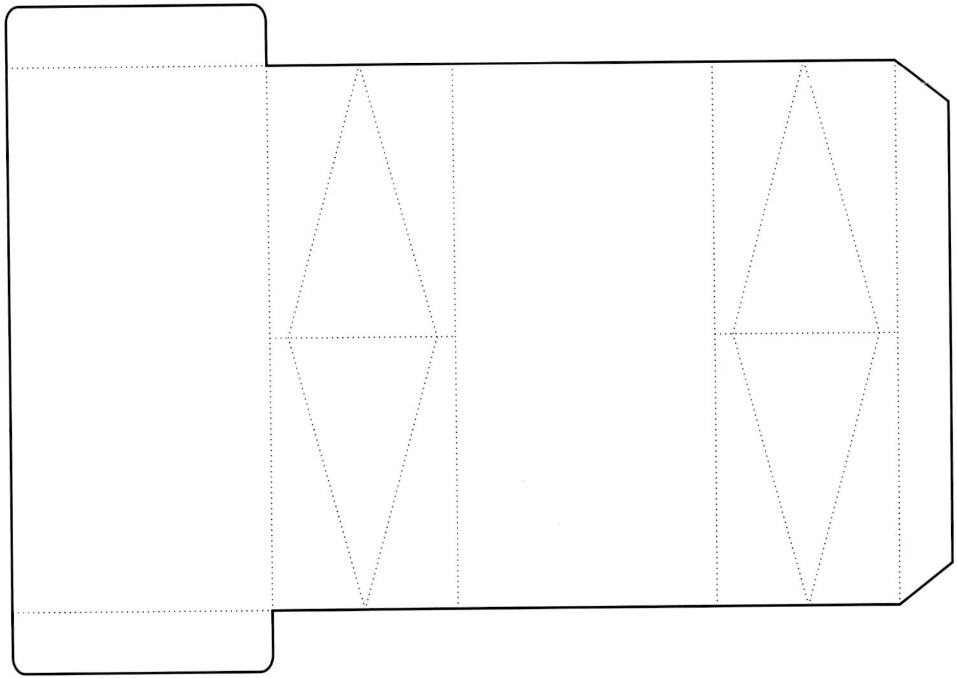

上海市著名商标
蟹派
CRAB
净含量：120 克

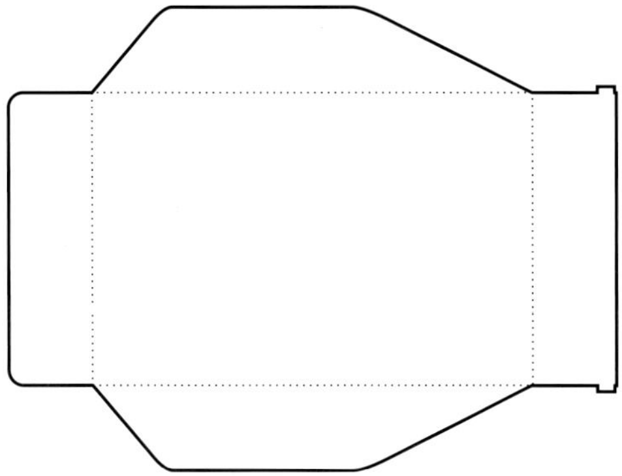

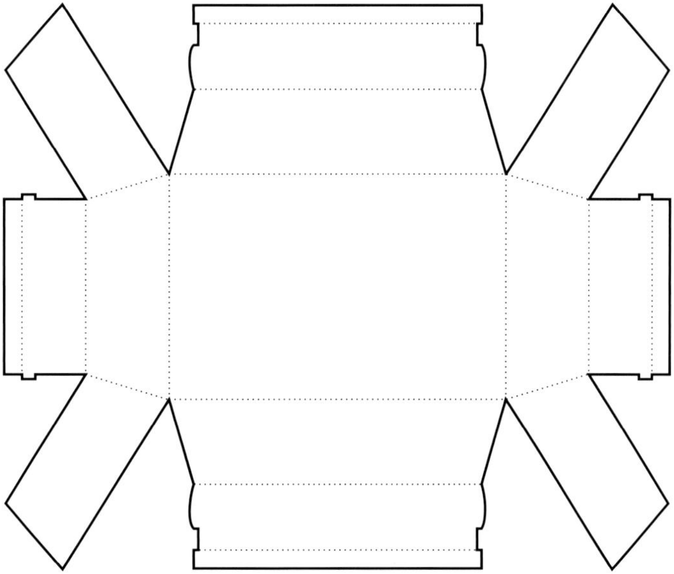

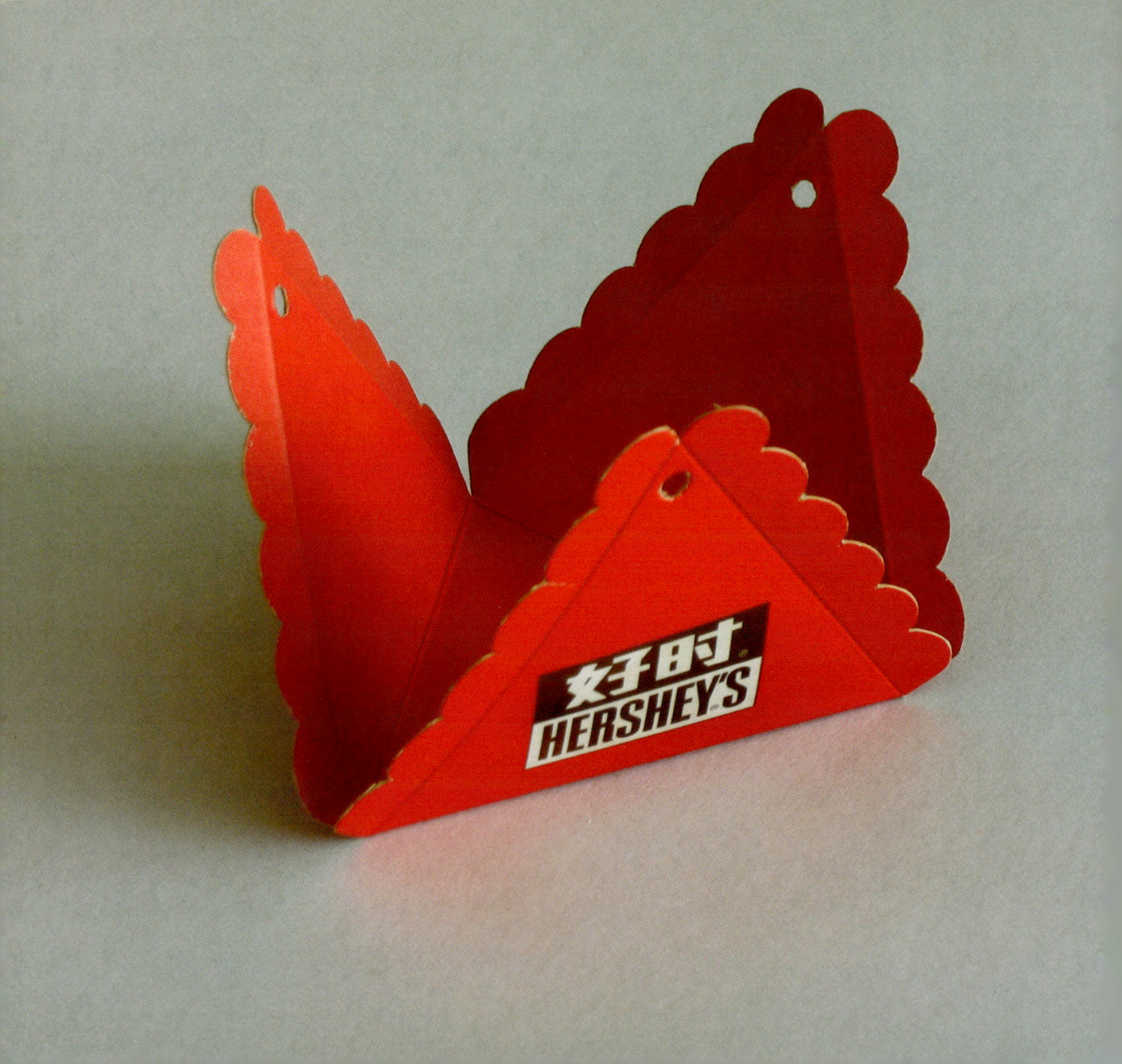
好时
HERSHEY'S

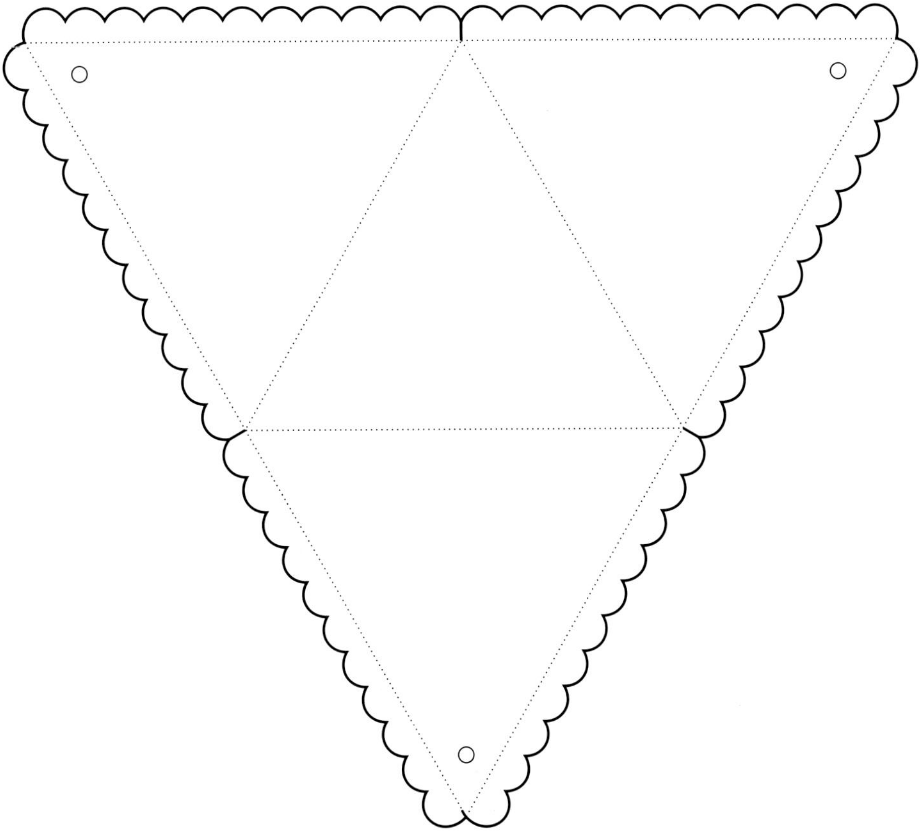

惠普笔记本电脑促销赠品
绚彩
甲壳虫
耳机
invent
绚彩
甲壳虫
耳机
惠普笔记本电脑促销赠品

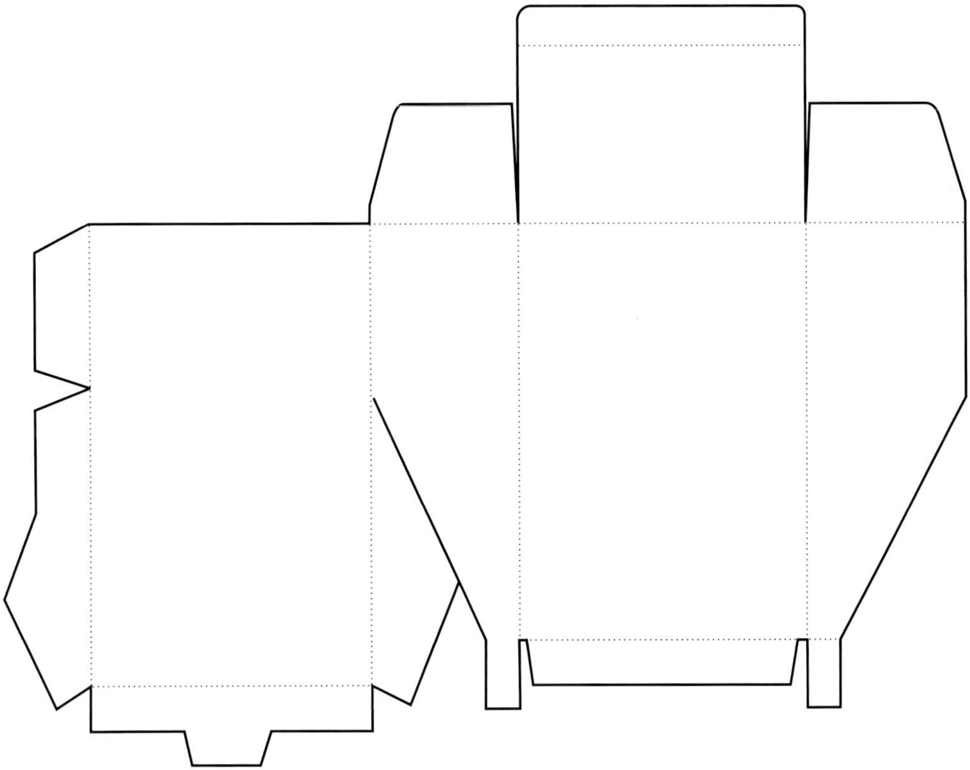

SHINE TIEN

廣合
ALLIANCE
BISCOTTI
Delicious!
百时可
一源自意大利的曲奇饼干
BISCOTTI
作为曲奇中的一大类
Biscotti(百时可)源于意大利
风靡欧美多年
一杯咖啡，几块Biscotti
尽享休闲时光

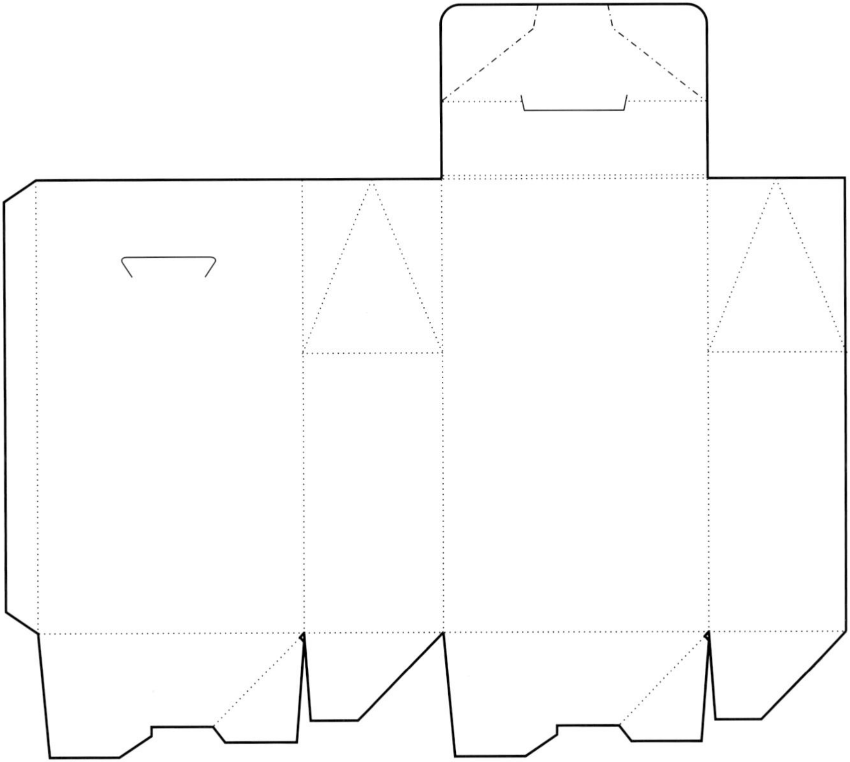

囍

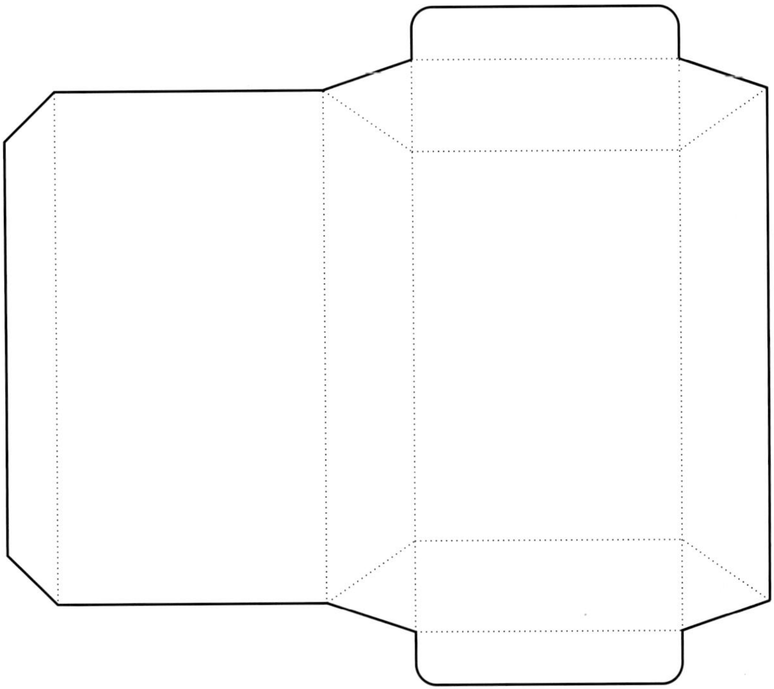

Swiss White Chocolate with Honey and Alm
TOBLERO
· of Switzerland ·

诗蒂
Shinetien
诗蒂
Shinetien
HAPPY

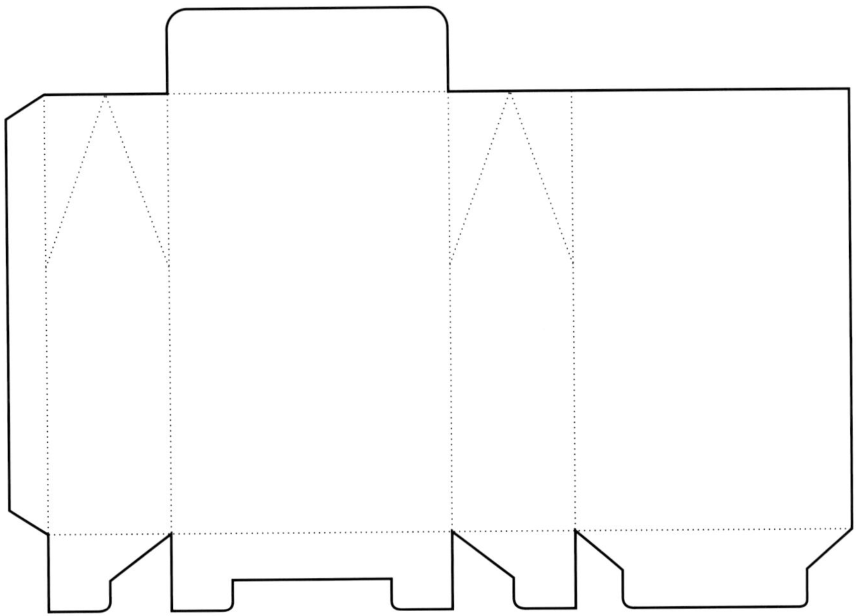

白板
豆酥
white bean
cake

Zàini
Caramelle ripiene ai gusti di frutta e creme
Grande
Assortimen
Taormina
Zàini

Shine Tien
Chocolate

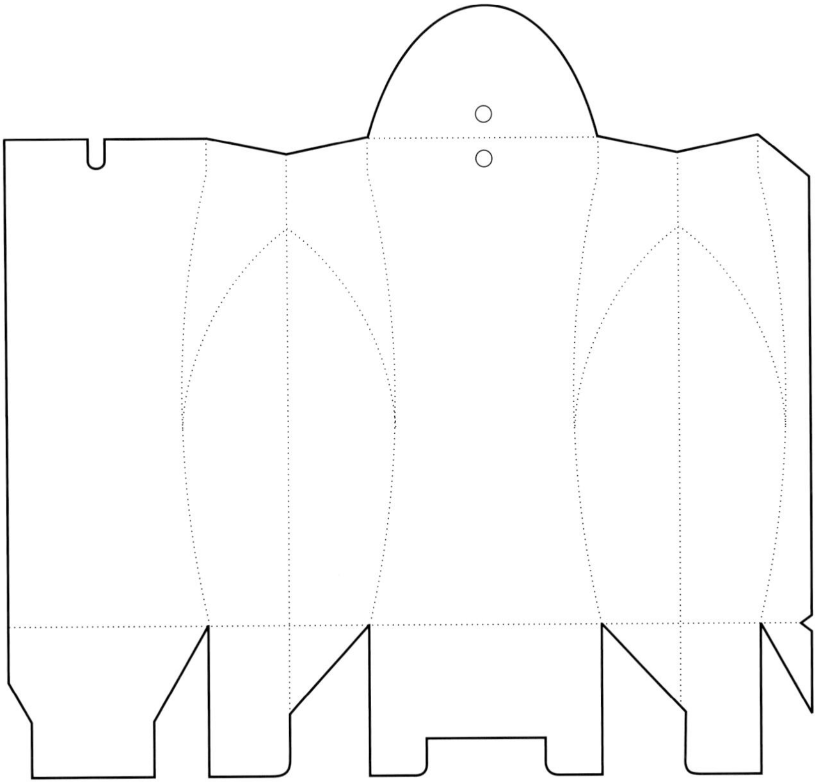

黑妞麻酥
Black
Sesame
Cake

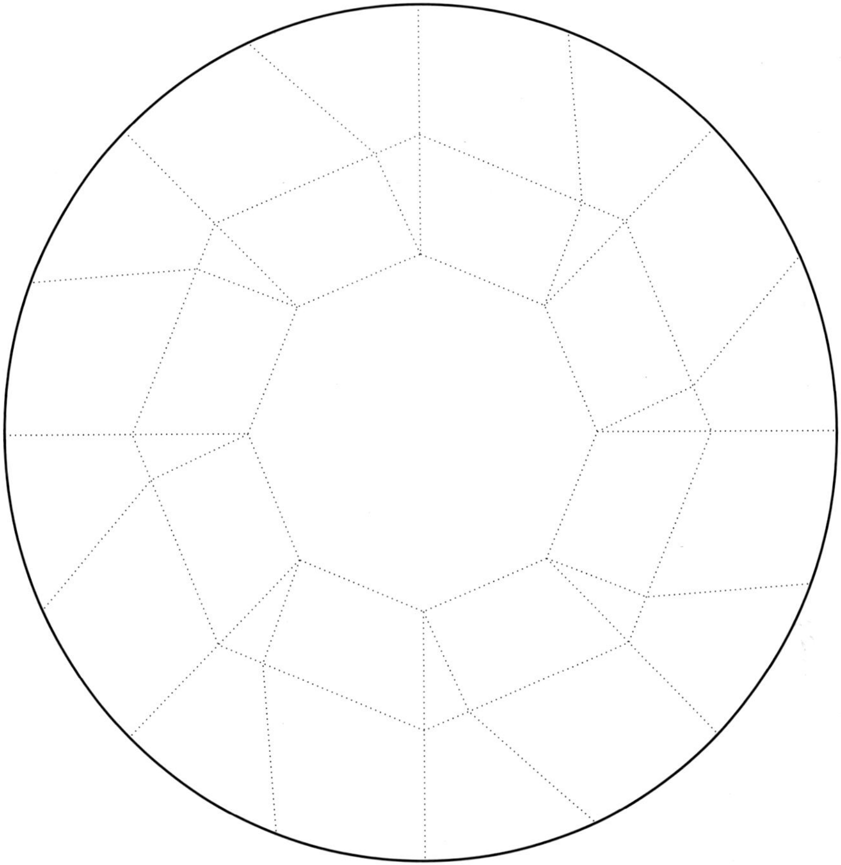

牛皮糖
gelled
confection

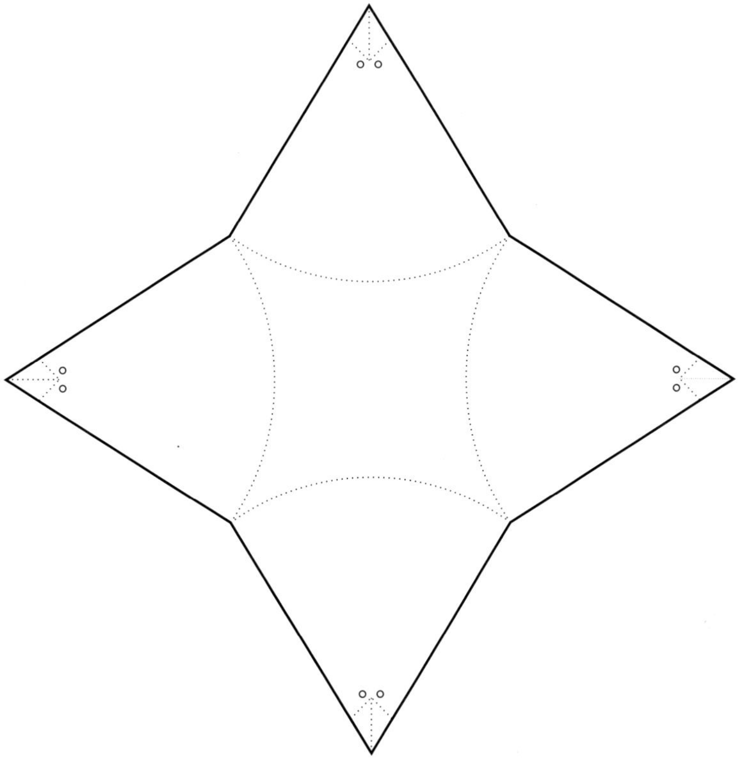

松仁
粽子糖
candy with pine nut

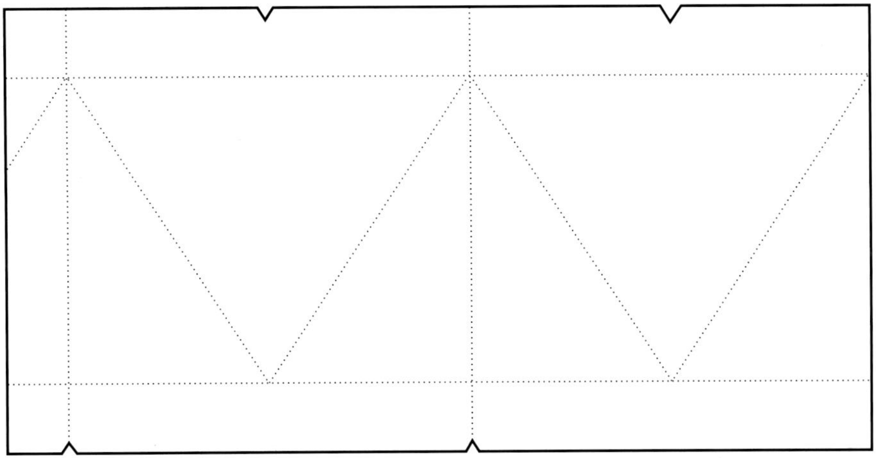

宝贝的健康日记
饮食会影响到孩子的健
康成长及智力发育，很多孩
子都不同程度会有挑食、
厌食、偏食及消化不良现
象。怎么样让孩子吃饭香
？是众多家长非常关心的
问题。青岛泽生堂"我要
吃饭香"开胃消食复合高
钙片含有丰富钙质、乳酸
锌、维生素C等，让您的孩
子在香香的草莓形钙片的
陪伴下健康成长每一天。

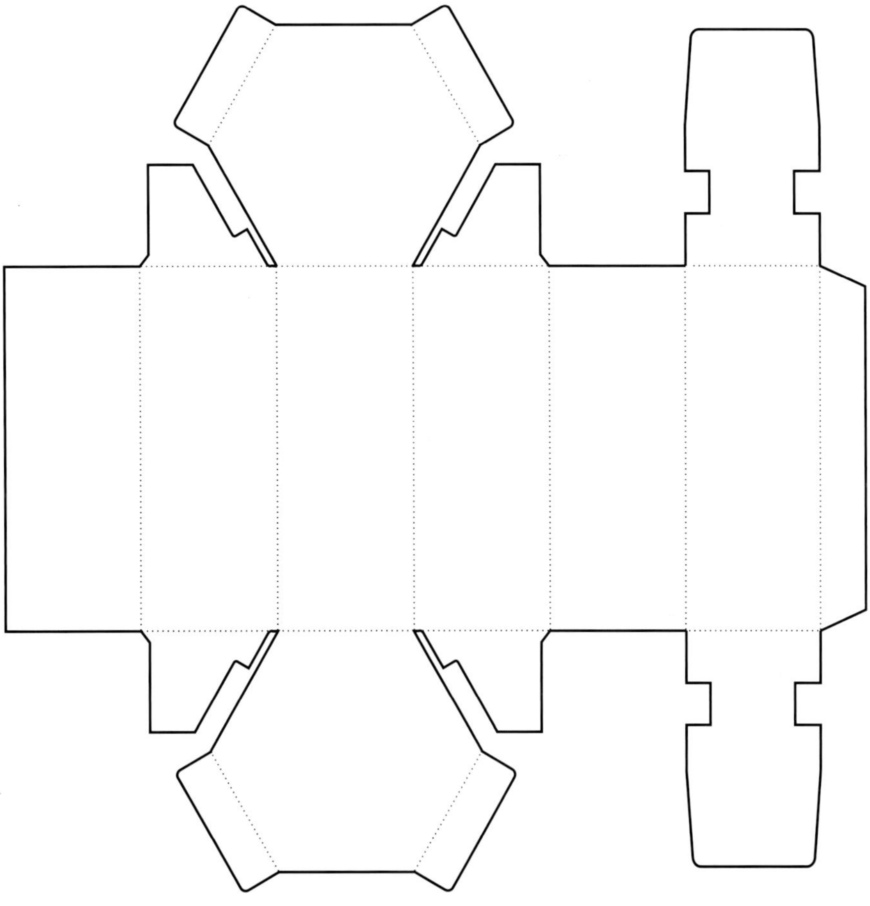

TIANTIAN®
天天星期五
天天
天天星期五
Friday
Lucky Day
蜜瓜果汁饮料
蜜瓜果汁饮料
净含量:230ml
蜜瓜果汁饮料
蜜瓜果汁饮料
净含量:230ml
It's Good Taste

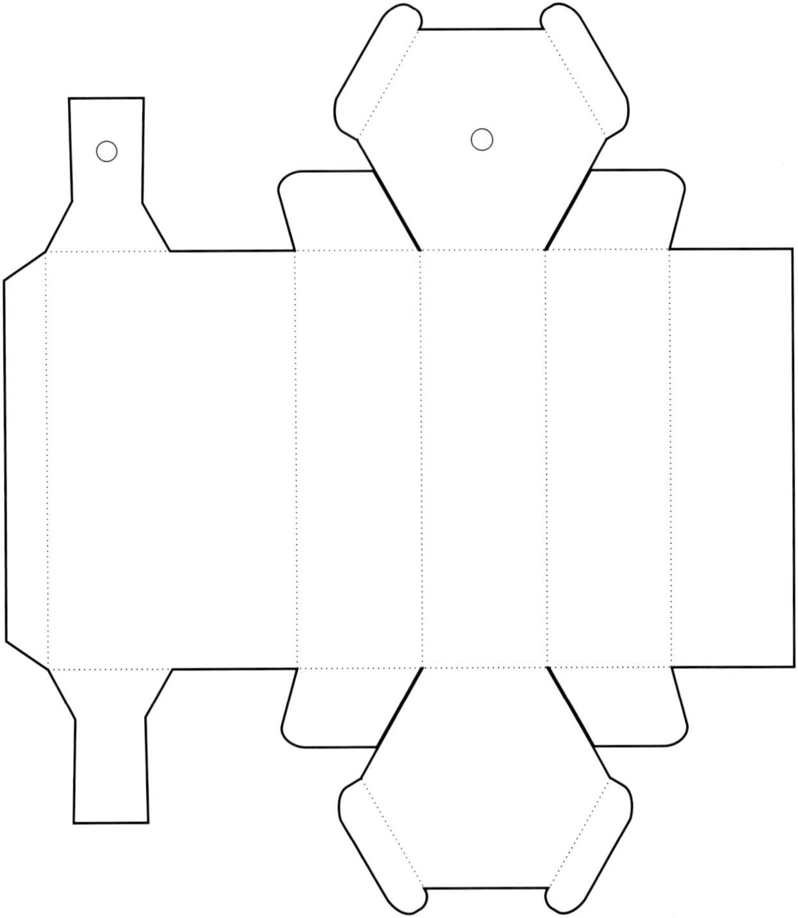

喜临门
XI LIN MEN
多味西梅
波罗梅 李 绿茶梅系列
喜临门
金玉满堂

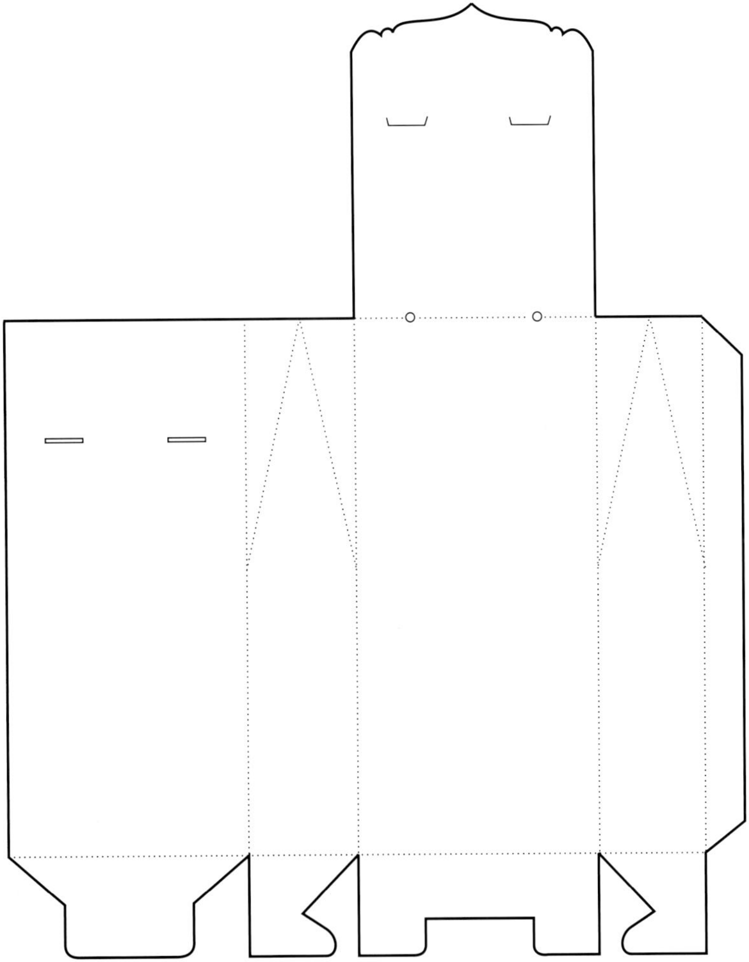

GREAT WALL
GRAPE WINE
GREAT WALLGRAPE WINE
GREAT WALLGRAPE WINE
GREAT WALLGRAPE WINE
GREAT WALLGRAPE WINE
GREAT WALL
GRAPE WINE

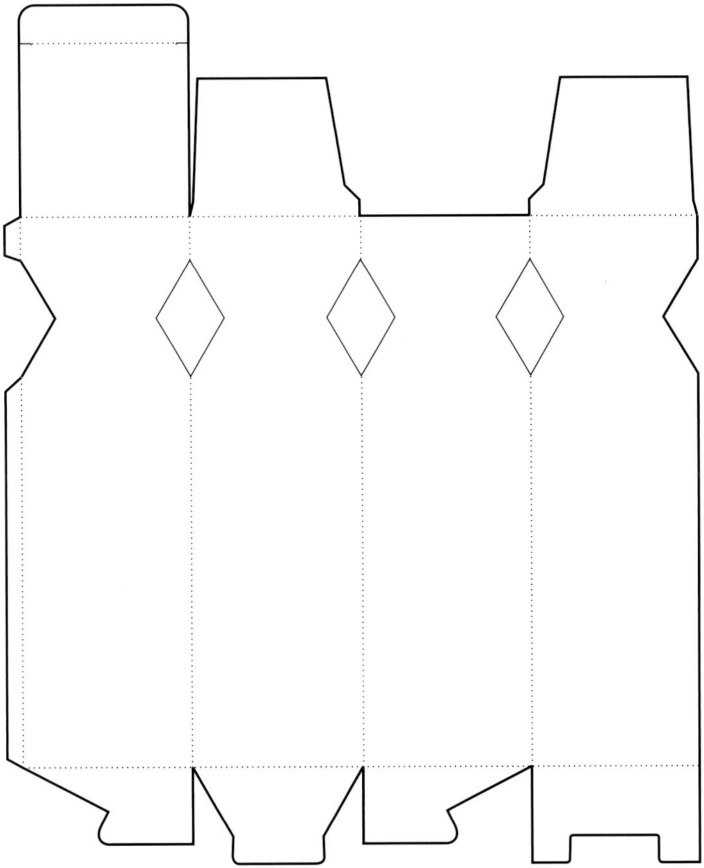

ShineTien

Famous chocolate
SHINETIEN
Shinetien
Shinetien

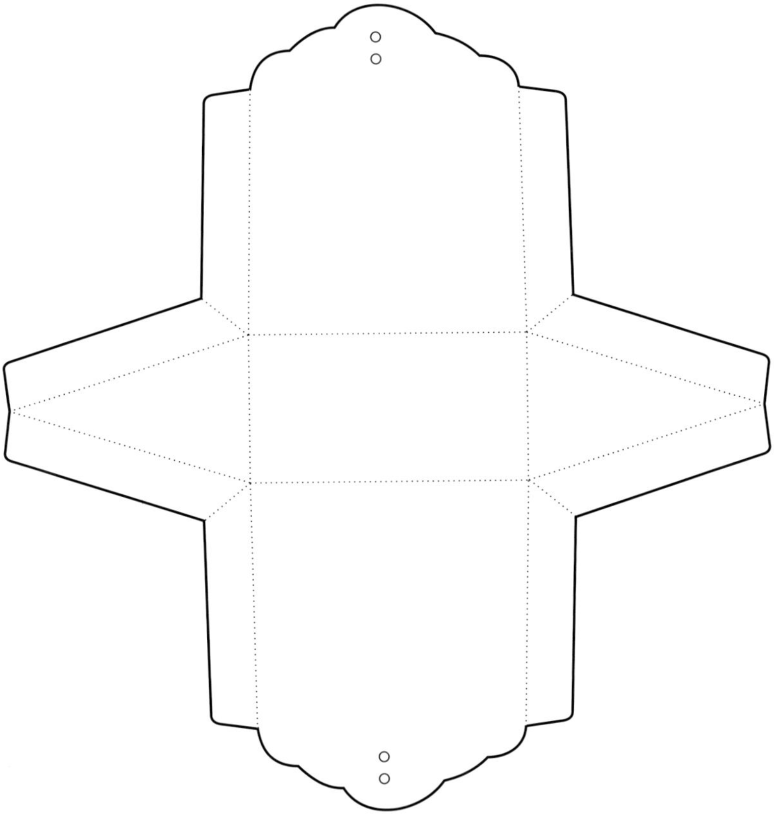

好时 HERSHEY'S
KISSES™
好时 巧克力
HERSHEY'S KISSES
百年 好合
好时 HERSHEY'S

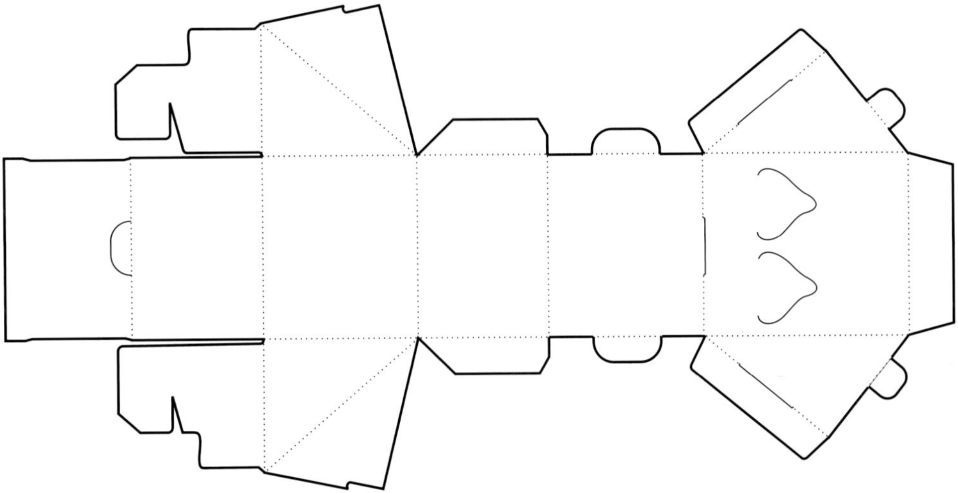

KFC
烹鸡美味
尽在肯德基

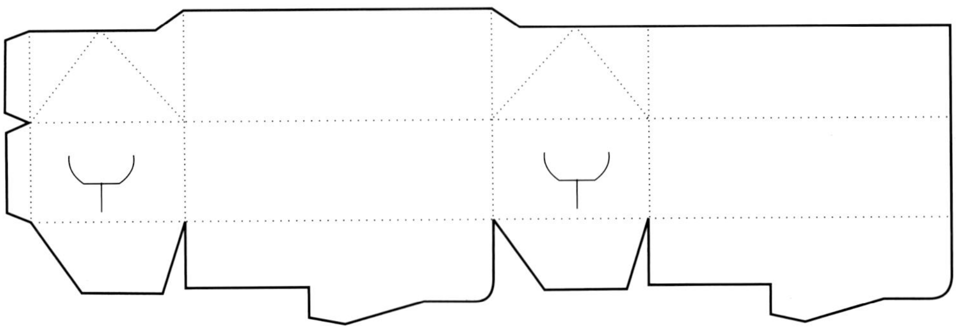

富贵吉祥
中秋月饼
VALUED AND EXQUISITE MID-AUTUMN GIFT!
MOON CAKE

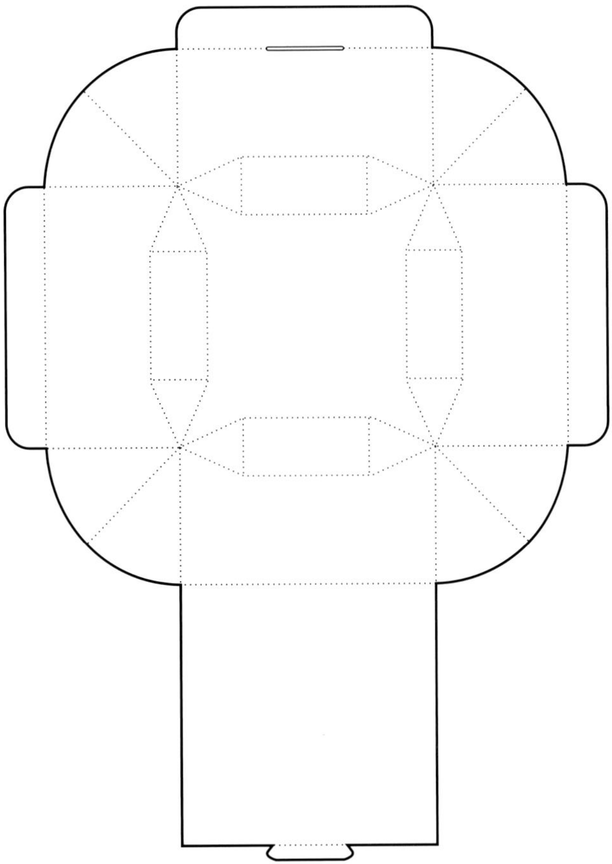

Côte d'Or
Since 1883
Sensations
Lait
INTENSE
Côte d'Or
Sensations
Lait INTENSE

TM
金威园
JINWEIYUAN
金威园 · 系列高級蜜餞
原味
西梅
PRUNES

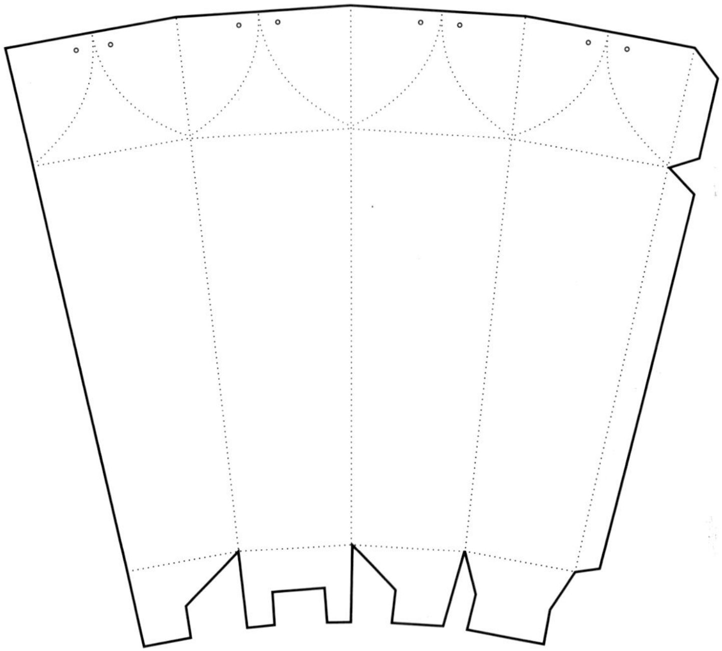

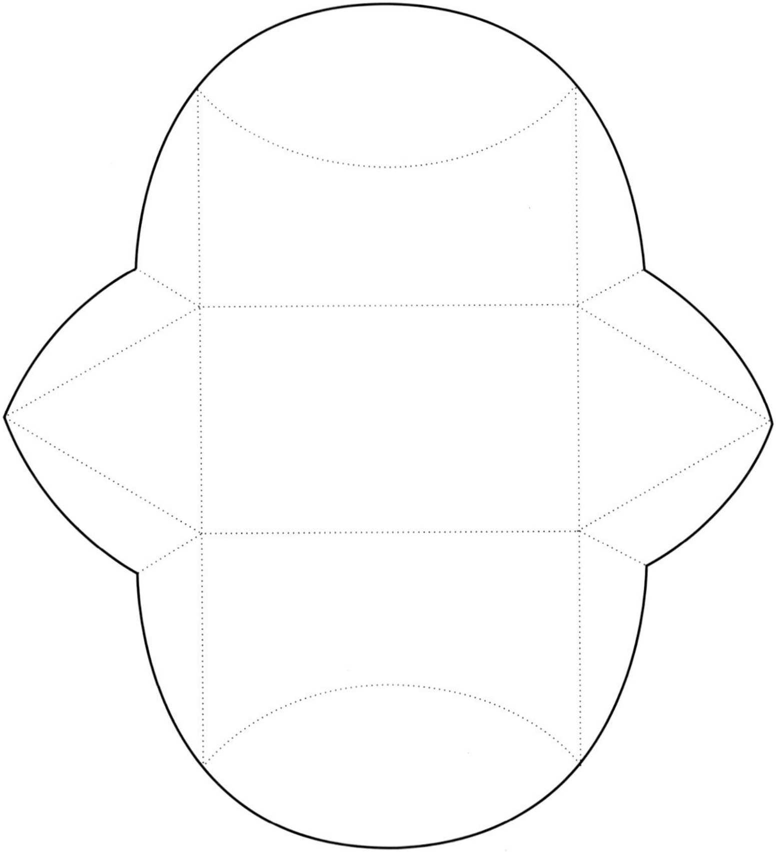

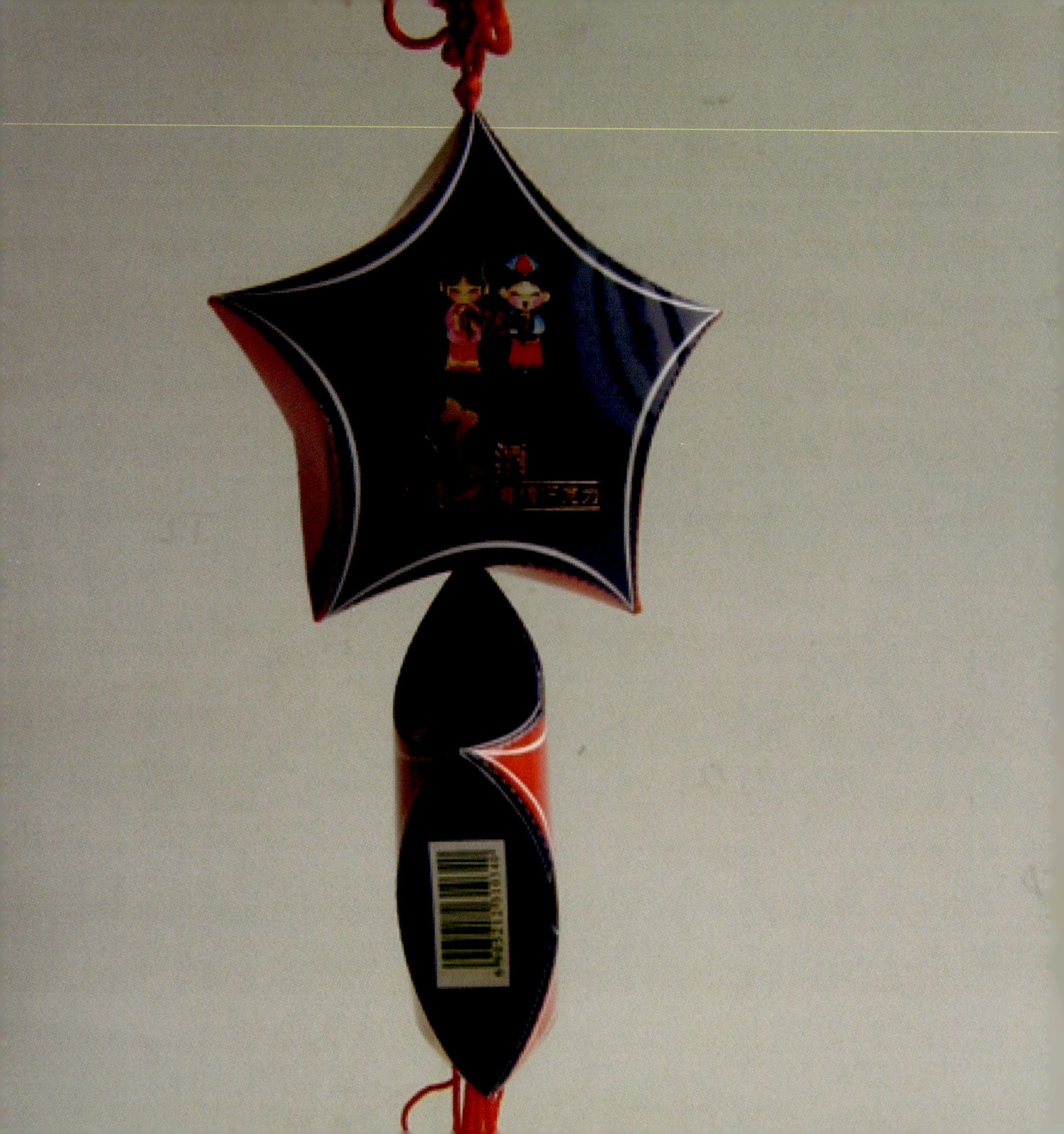

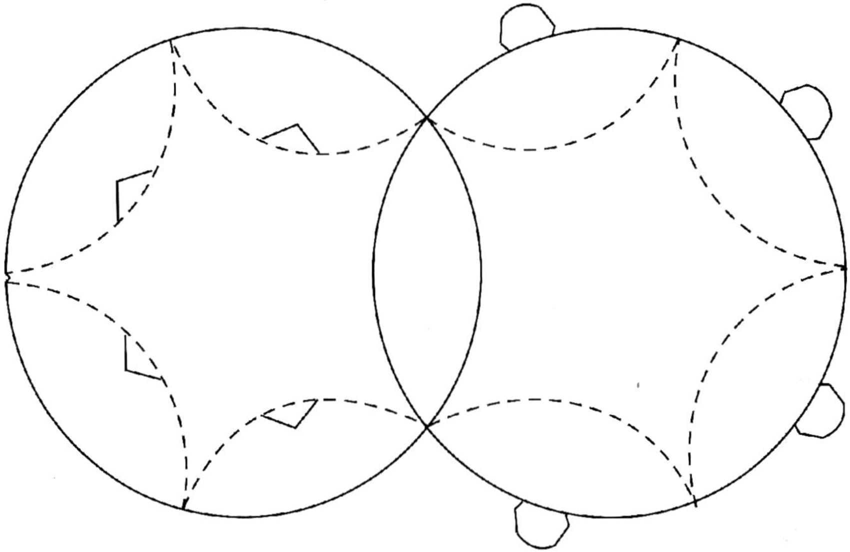

囍

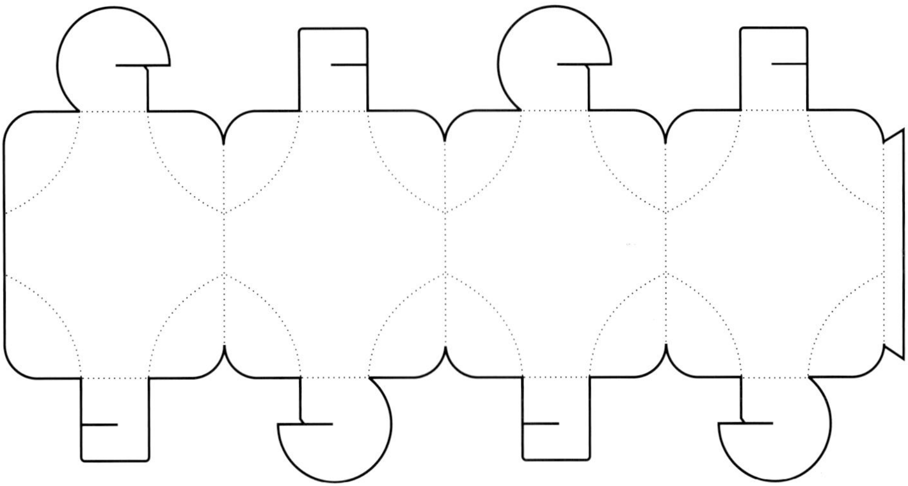

虞姬神
江苏省明星企业
虞姬神
江苏虞姬神酒业有限公司
JIANGSUYUJISHENJIUYEYOUXIANGONGSI

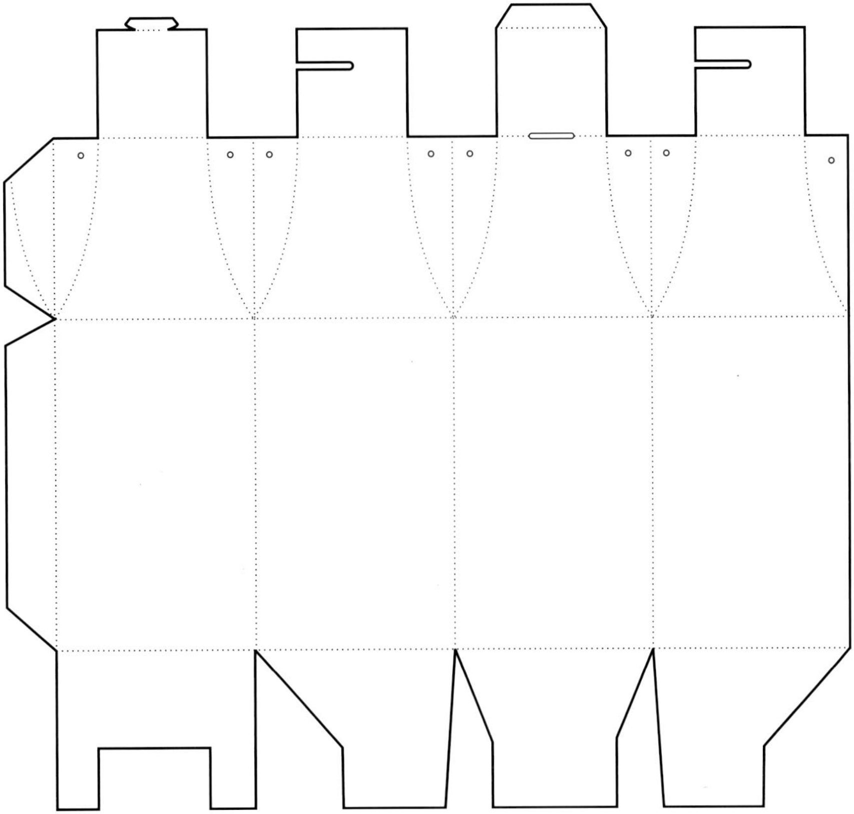

德
德芙
Dove

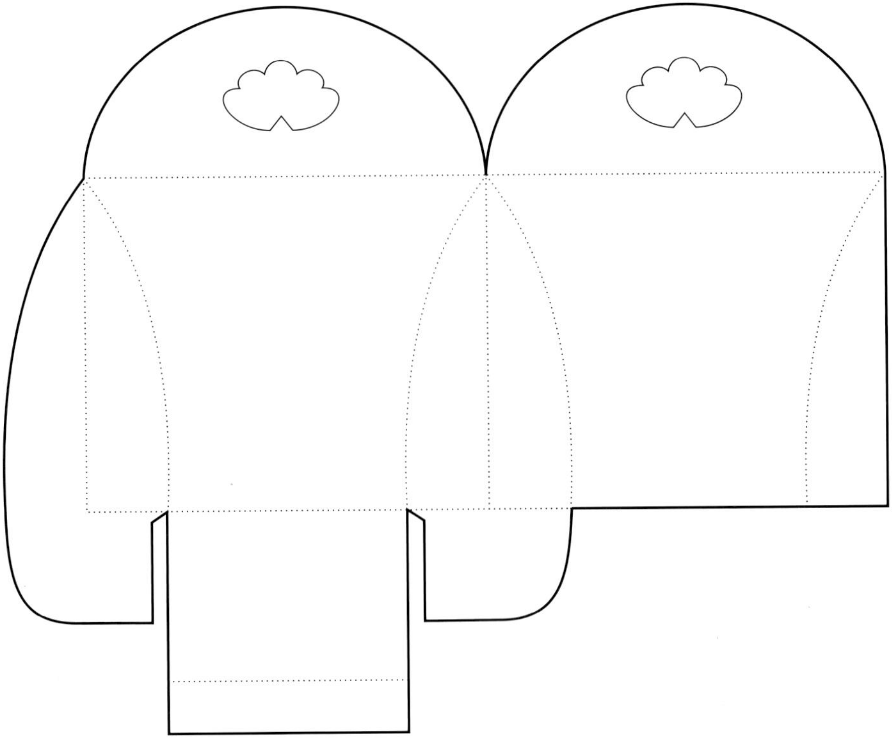

Welcome
wedding party!
Shinetien

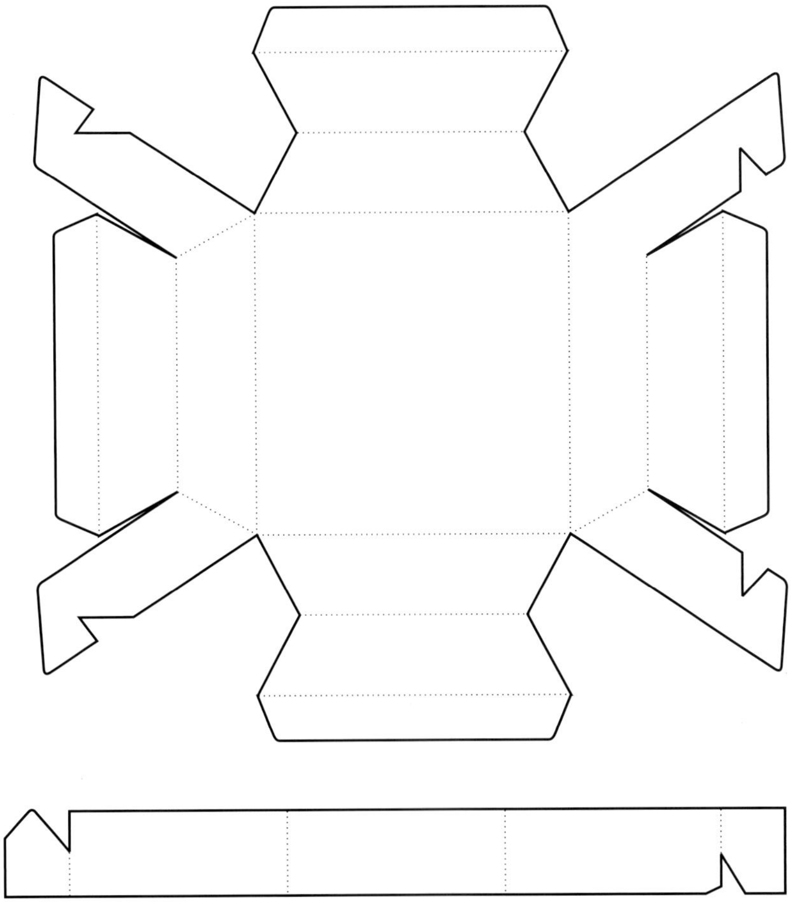

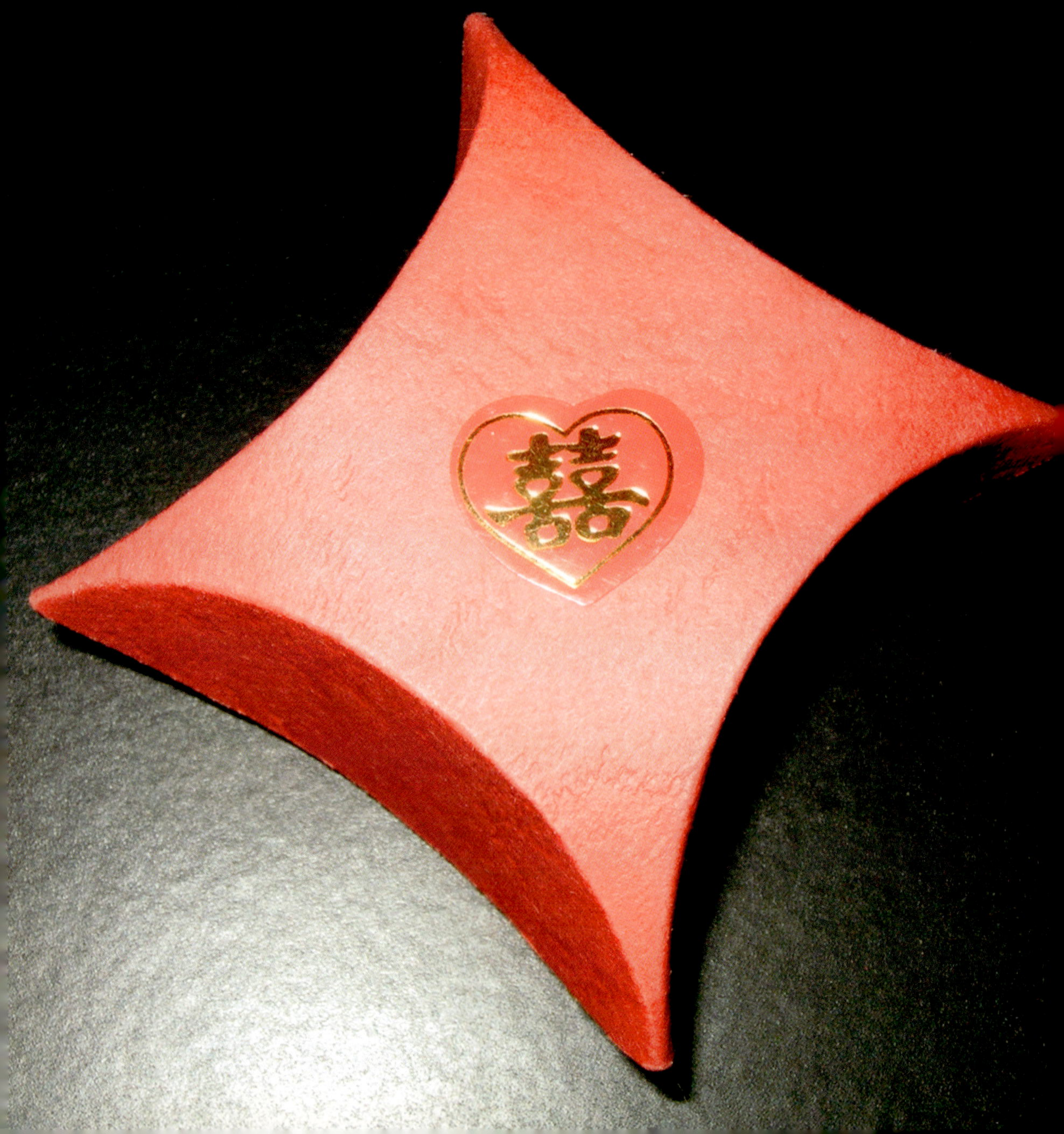
囍

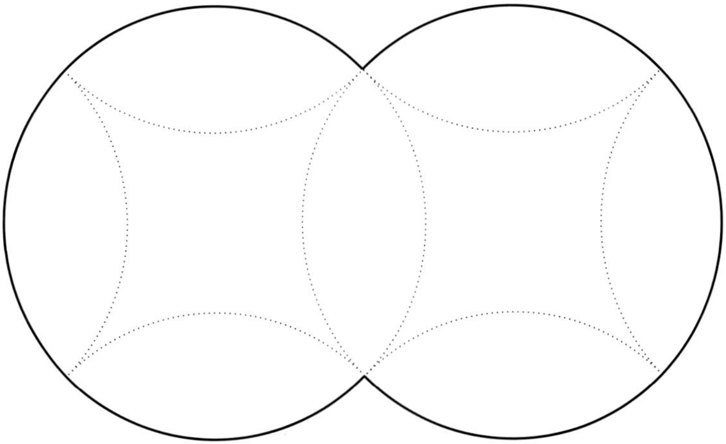

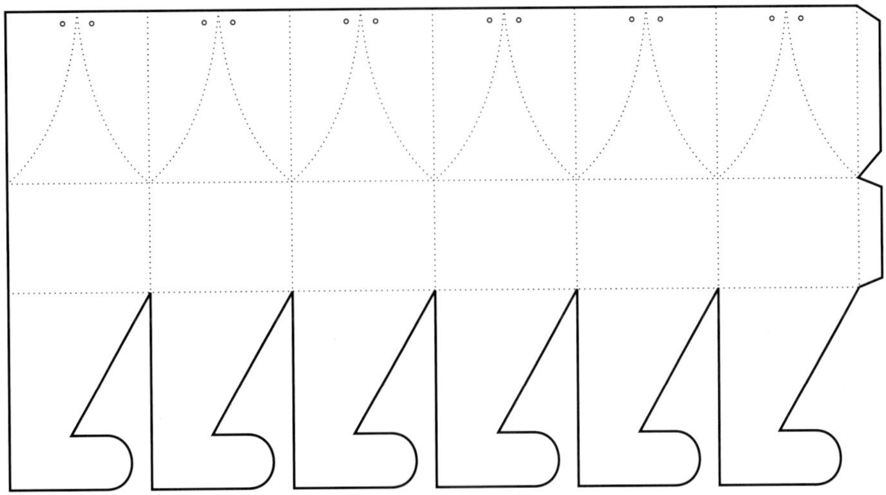

中秋月饼

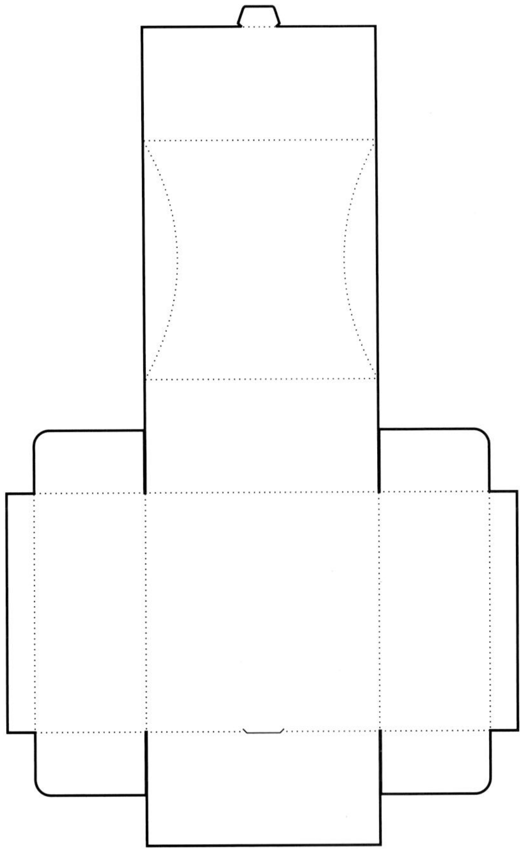

サワデー
爽花蕾
日本技术
芬芳新鲜
香囊
薰衣草香型
替换装
KOBAYASHI

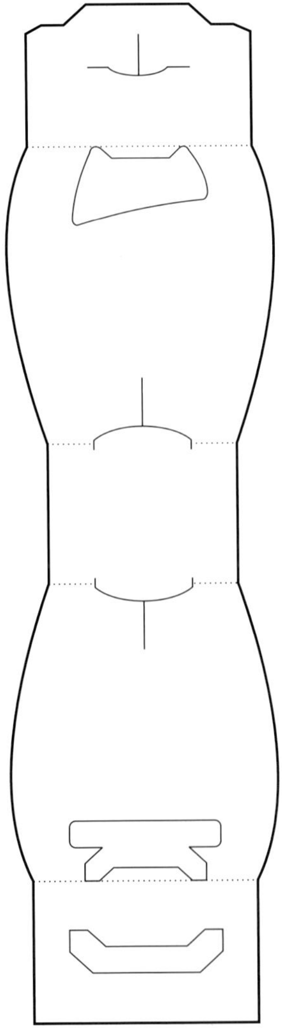

畲茶

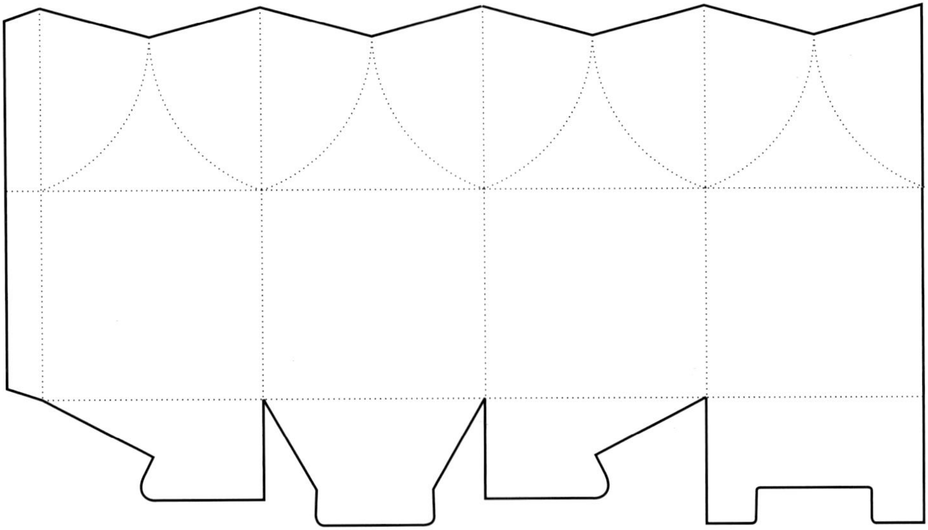

Shinetien
ShineTien

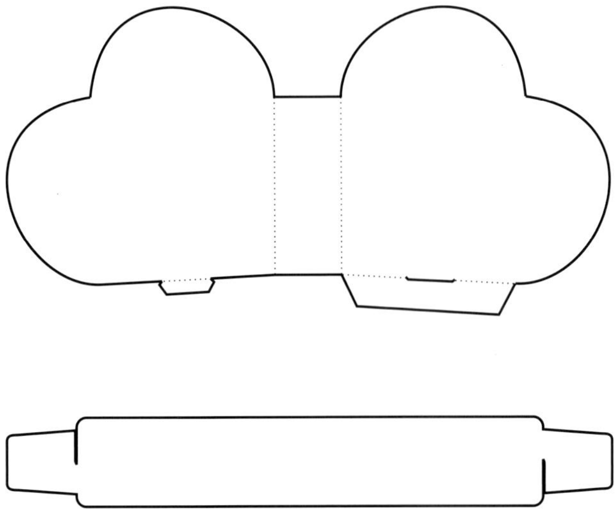

UHA悠哈
特浓口味

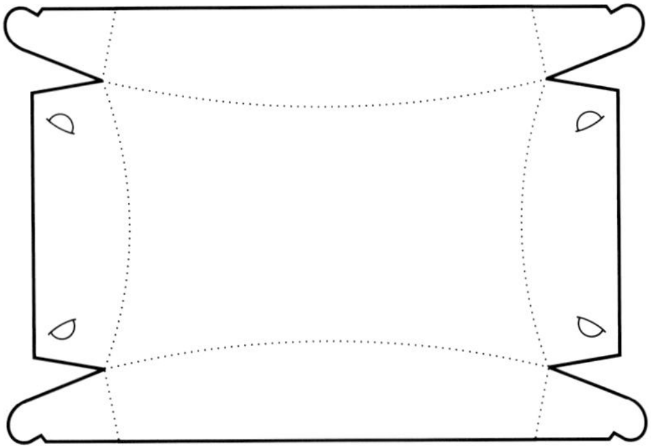

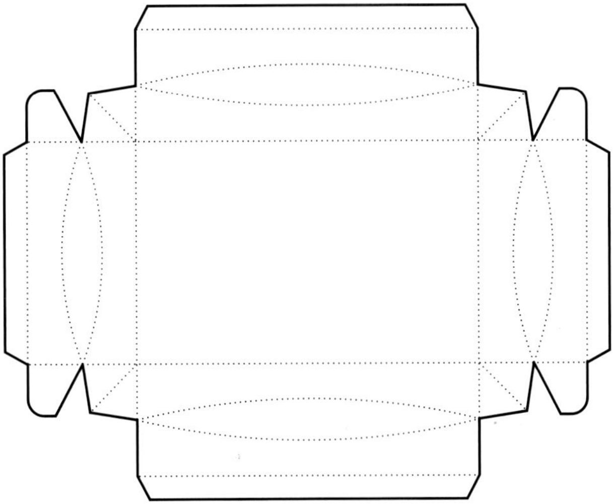

椒盐酥
Five nuts
salt cake

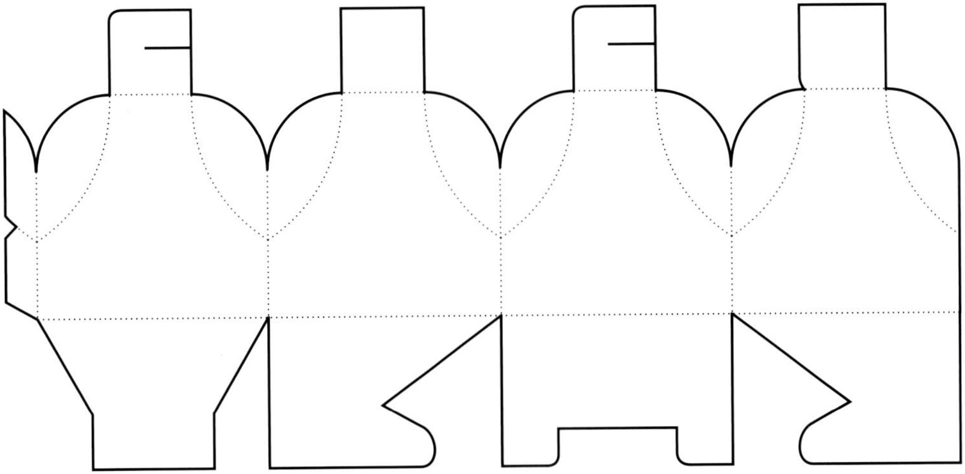

6块
上校鸡块
Kentucky Nuggets™
Nuggets
KFC®
6块

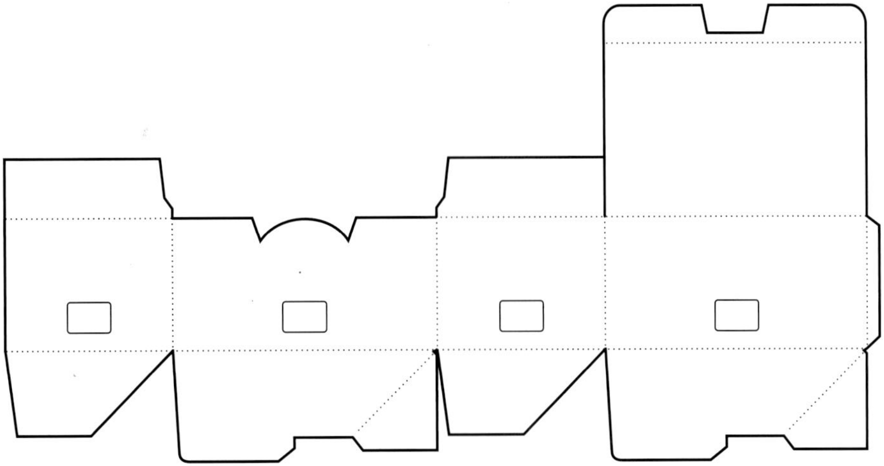

Sweet Dessert
甜蜜点心

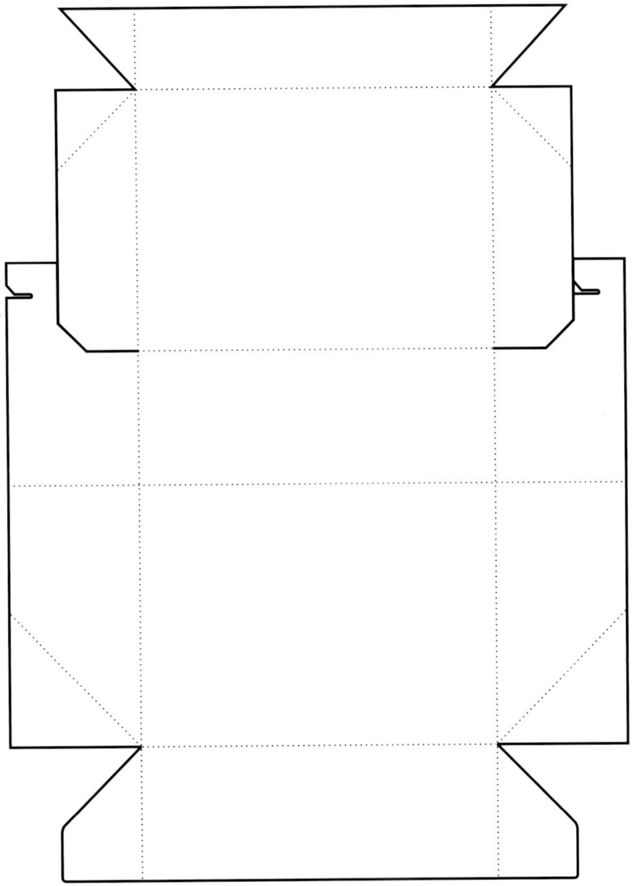

MOON CAKE

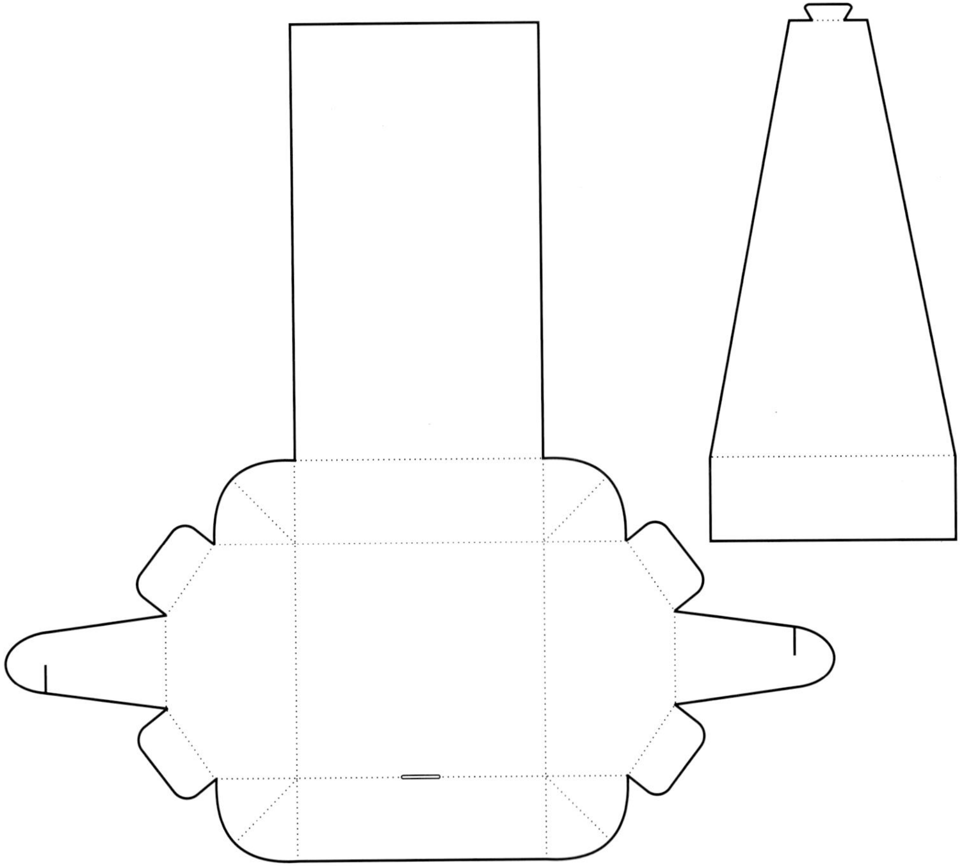

Exciting PANORAMA function
OLYMPUS
xD-Picture Card
Only Olympus xD-Picture Card has exciting Panorama function when used with Olympus digital camera. Please make sure to purchase Olympus branded xD-Picture Card for more fun.
OLYMPUS
Optional Accessories
External flash
Flash bracket cable
Flash bracket
USB Reader/Writer
xD-Picture Card
Extension lenses
Battery charger
Digital Cameras
Compatibility depending on camera model
Underwater cases
Leather cases
Printed in China

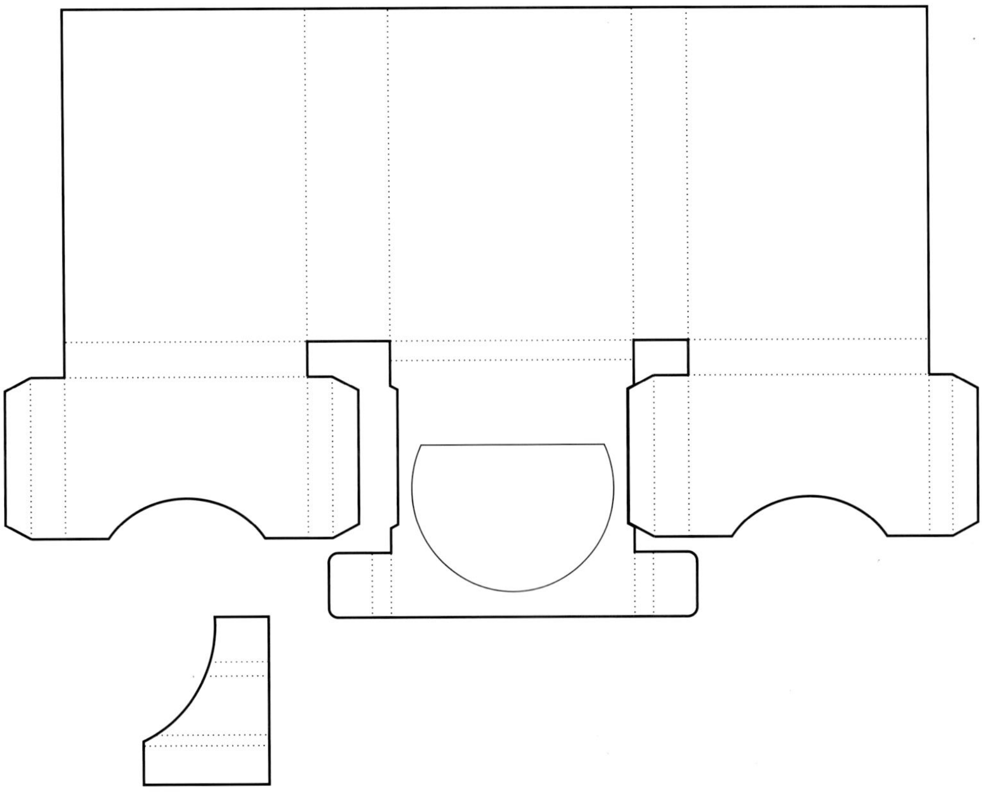

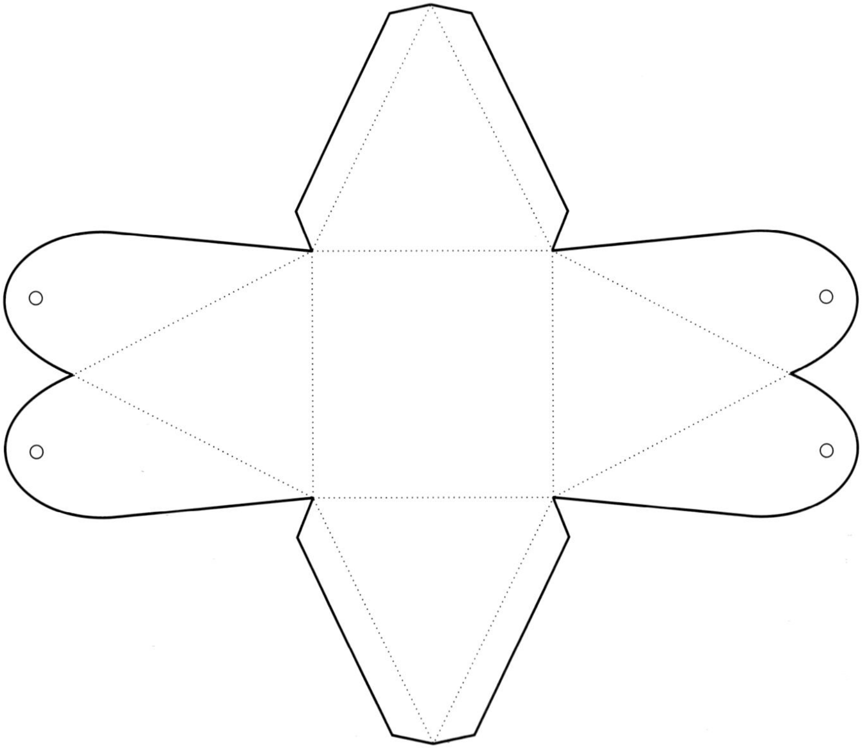

HUA JIA MARRIANG SHOP

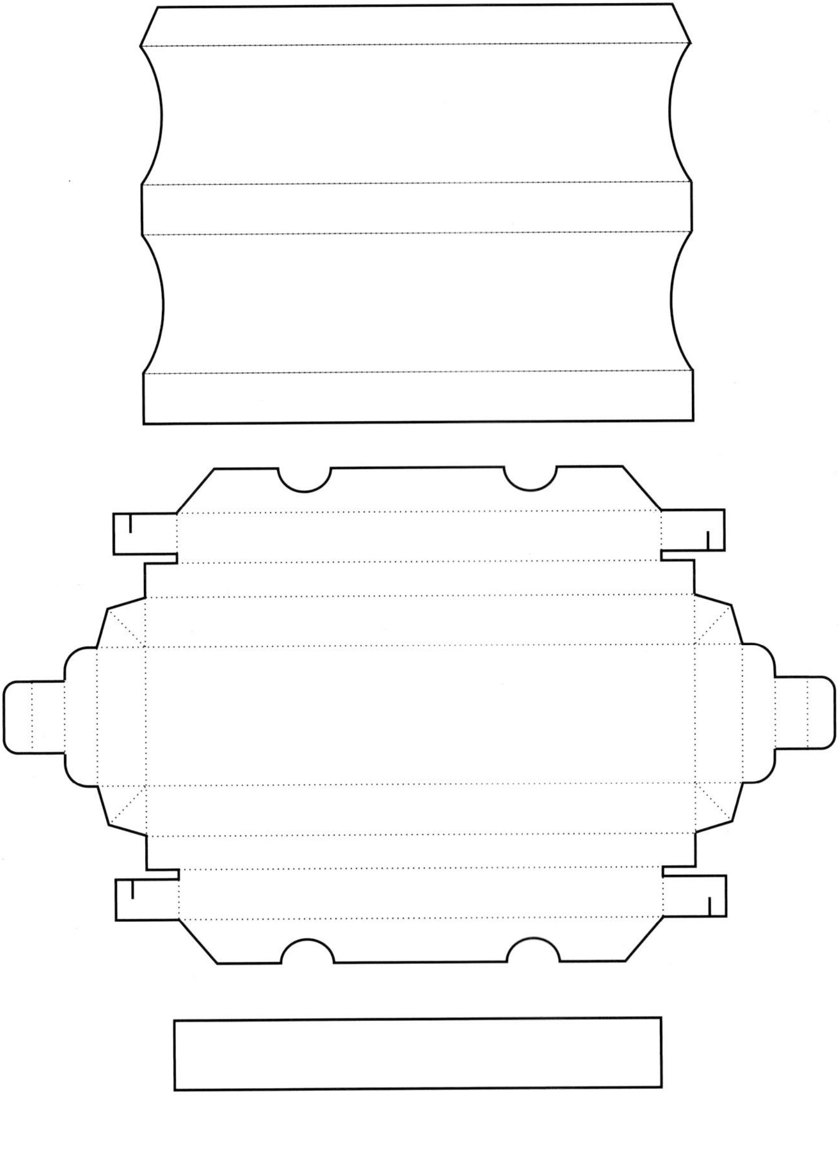

Pioneer 先锋
16X16
全能刻录机
16X DVD±R
8X DVD±R
6X DVD±R DL
8X DVD-RW
5X DVD+RW
40X CD-RAM
32X CD-R
32X CD-RW
(米白色)
Super Multi Drive
DVR-110AXL
全能刻录机
ATAPI
www.pioneer-dvd.com.cn

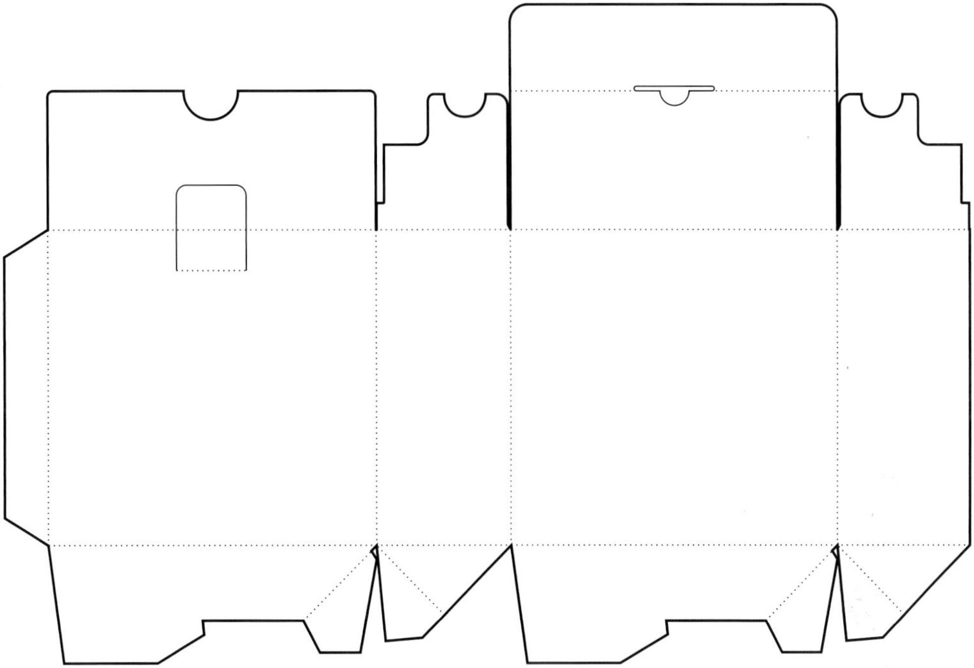

Nikon
尼康
尼康数码相机
COOLPIX 4800
ZOOM NIKKOR ED
6-50mm 1:2.7-4.4
400万！有效像素
4.0
8.3倍变焦
ED尼克尔镜头
8.3x
15种场景模式
15
PictureProject
Nikon
尼康数码相机
COOLPIX 4800
量子全面服务标志

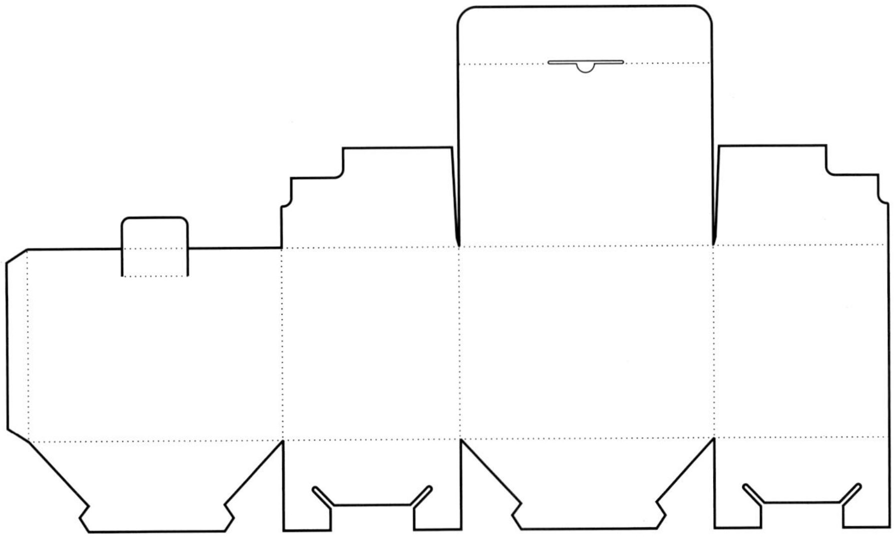

SEASON'S
Greetings
Christmas Cards

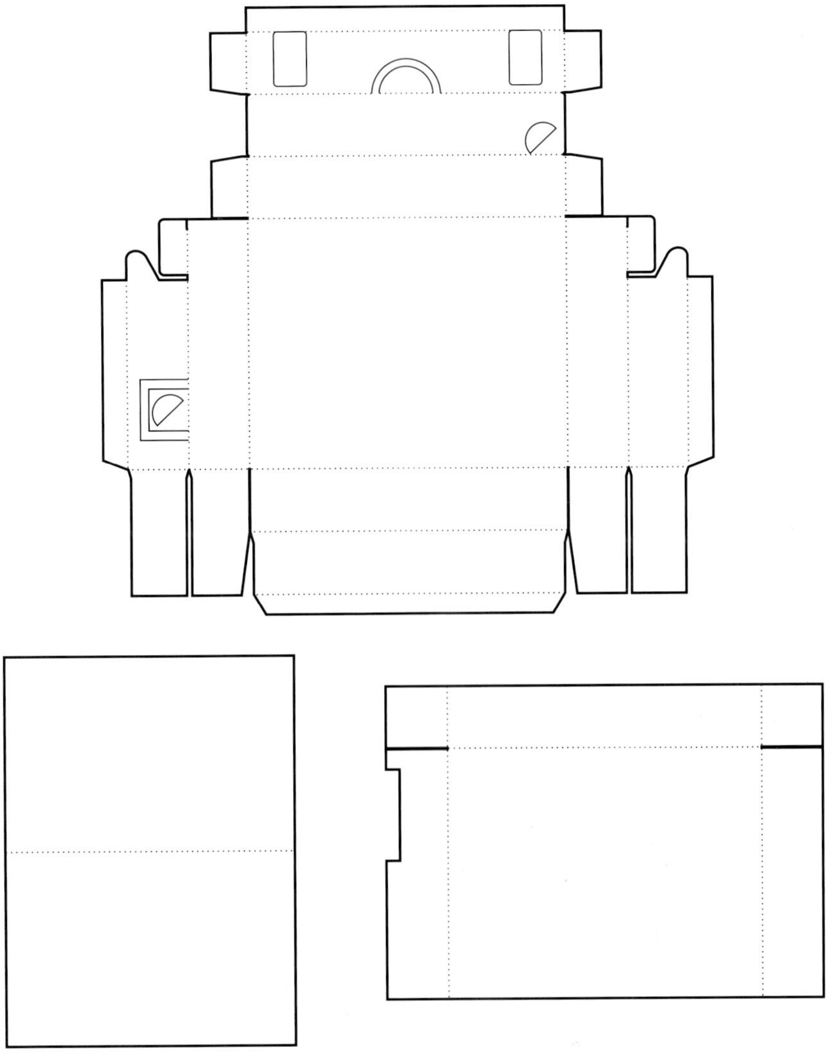

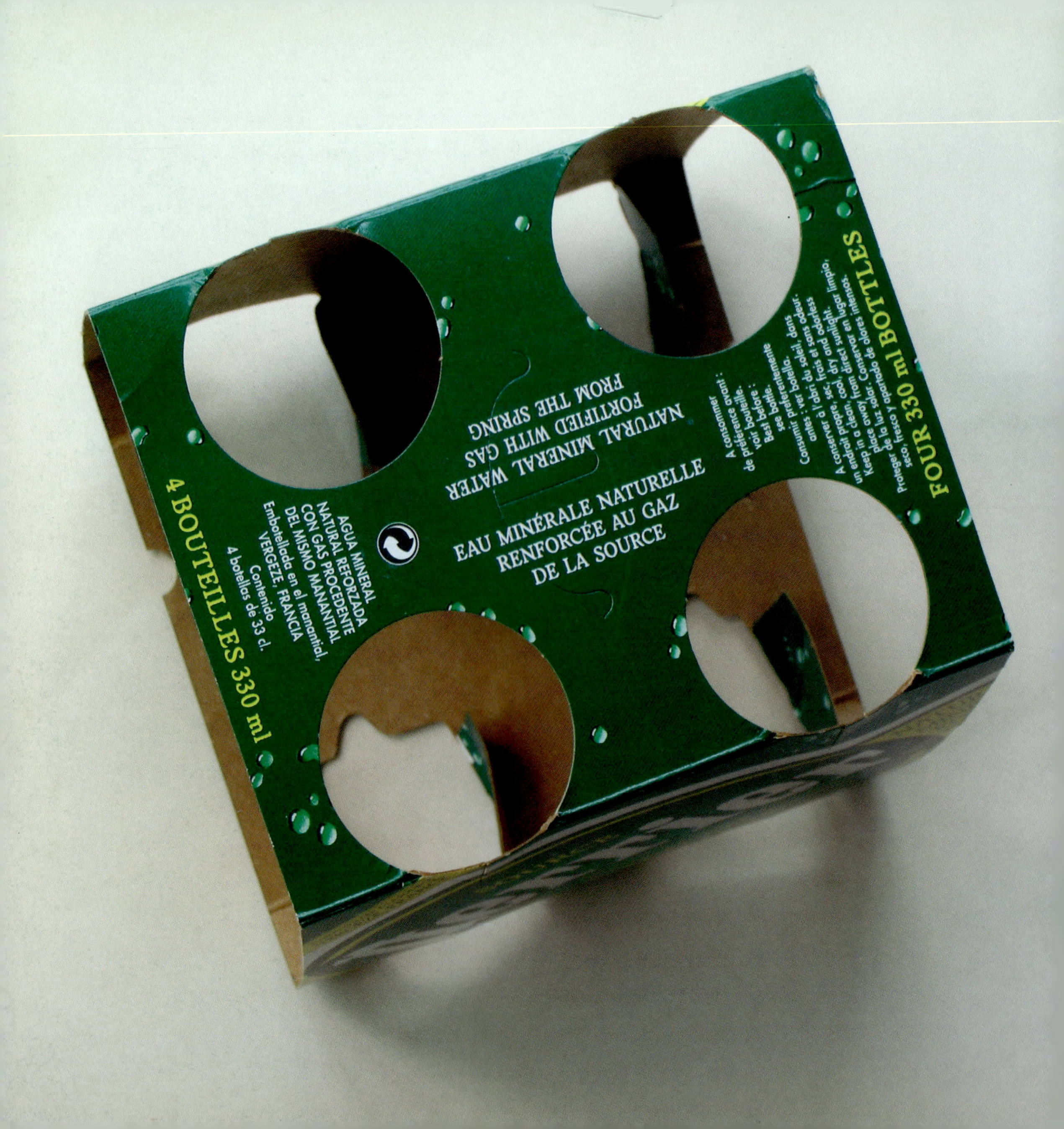
EAU MINÉRALE NATURELLE
RENFORCÉE AU GAZ
DE LA SOURCE
NATURAL MINERAL WATER
FORTIFIED WITH GAS
FROM THE SPRING
AGUA MINERAL
NATURAL REFORZADA
CON GAS PROCEDENTE
DEL MISMO MANANTIAL
Embotellada en el manantial,
VERGEZE, FRANCIA
Contenido
4 botellas de 33 cl.
4 BOUTEILLES 330 ml
FOUR 330 ml BOTTLES

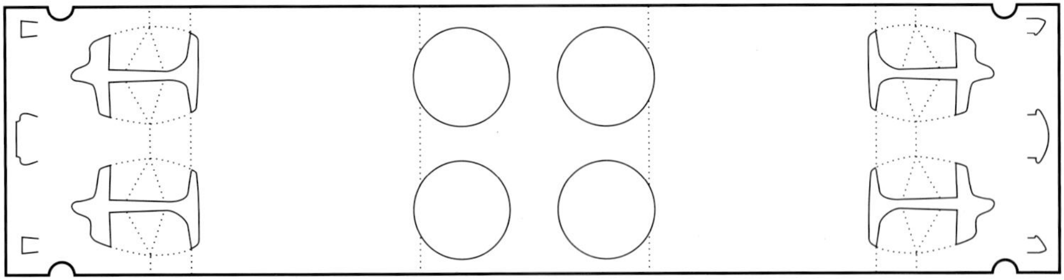

Corona Extra
Imported Beer From Mexico
BIER
CERVEZA
BIERE

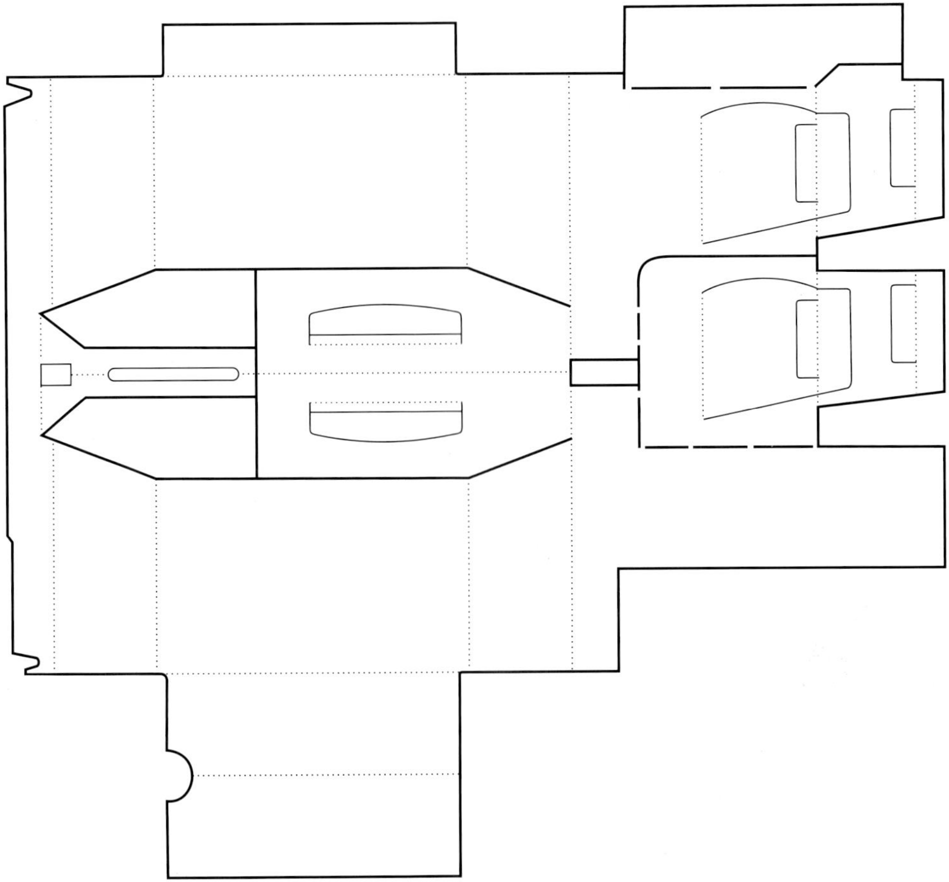

Acknowledgment

Thanks to all my friends who helped with the creation of this book.

My concept became a reality as a result of your kind assistance.